EXPLORING SPACE:

OPENING NEW FRONTIERS

Dr. Al Koller

COPYRIGHT © 2016, A. KOLLER, JR.
All rights reserved. No part of this book may be reproduced without the written consent of the copyright holder

Library of Congress Control Number: 2016917577

ISBN: 978-0-9668570-1-6
e3 Company
Titusville, Florida
http://www.e3company.com

TABLE OF CONTENTS

	Page
Foreword	2
Dedications	3
A Place of Canes and Reeds	4
Cape Canaveral and The Eastern Range	7
Early Missile Launches	9-17
Explorer 1 – First Satellite	18
First Seven Astronauts	20
Mercury Program	23-27
Gemini Program	28
Air Force Titan Program	29-30
Apollo Program	31-35
Skylab Program	35
Space Shuttle Program	36-40
Evolved Expendable Launch Program	41
Constellation Program	42
International Space Station	42
Cape Canaveral Spaceport Today	43
ULA – Atlas V, Delta IV	44
Boeing X-37B	45
SpaceX Falcon 1, Falcon 9, Dragon Capsule	46
Boeing CST-100 Starliner	47
Sierra Nevada Dream Chaser	48
Lockheed Martin Orion Capsule	48
The Future of American Space Exploration	50
NASA Space Launch System (SLS)	50
ULA Atlas V, Man-Rating Complex 41	50
SpaceX Falcon 9 Heavy, Complex 39A	50
Blue Origin New Shepard, BE-4 Engine, Complex 36	51
Moon Express Google Xprize, Complexes 17 & 18	51
OneWeb Satellite Manufacturing Plant	51
YOU CAN JOIN IN EXPLORING SPACE	52
Why It Matters Where You Launch	53
What It's Like Living in Space	54
How You Can Be A Part Of All This	55-57
Web Links for Additional Information	58
About the Florida Spaceport	59
About the Author	59

FOREWORD

"Exploring Space: Opening New Frontiers" is all about space exploration, space history, and space pioneers – **past, present, and future.**

It covers more than **500 years** and includes the early history of **Cape Canaveral** and the **Air Force Eastern Test Range** from the first launches there to the current activities at the **Cape Canaveral Air Force Station** and **NASA's Kennedy Space Center.**

Its goal is to highlight past achievements in space; current programs, missions and flights; and upcoming activities to describe where we've been, what's happening now, and where we are headed in the future.

**Cape Canaveral Light – Original and replacement. The old 1848 brick tower next to the 1868 iron tower with a white lantern room, before the latter was moved in 1893.
https://en.wikipedia.org/wike/Cape_Canaveral_Light.**

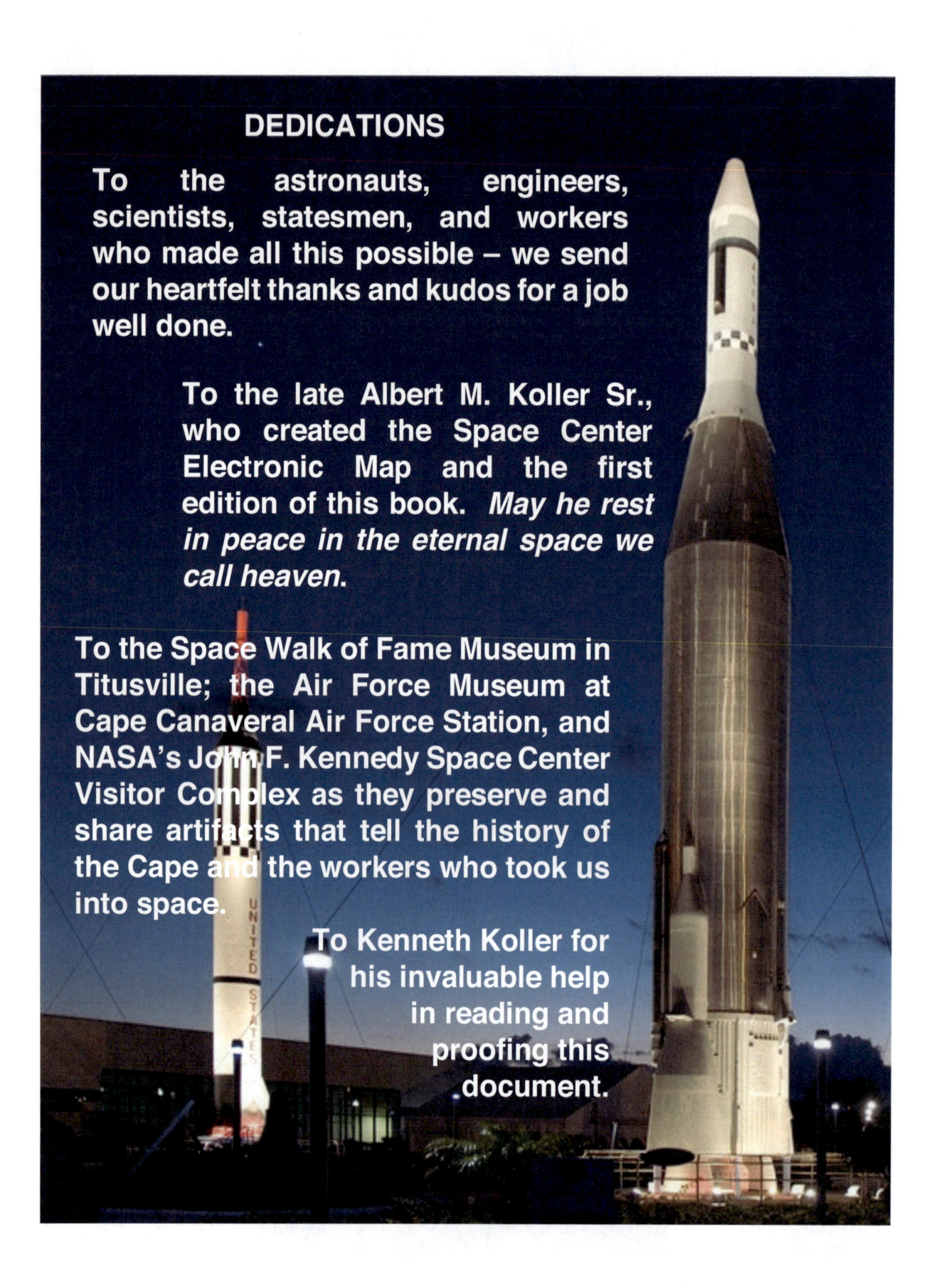

DEDICATIONS

To the astronauts, engineers, scientists, statesmen, and workers who made all this possible – we send our heartfelt thanks and kudos for a job well done.

To the late Albert M. Koller Sr., who created the Space Center Electronic Map and the first edition of this book. *May he rest in peace in the eternal space we call heaven.*

To the Space Walk of Fame Museum in Titusville; the Air Force Museum at Cape Canaveral Air Force Station, and NASA's John F. Kennedy Space Center Visitor Complex as they preserve and share artifacts that tell the history of the Cape and the workers who took us into space.

To Kenneth Koller for his invaluable help in reading and proofing this document.

THE CAPE -- A PLACE OF CANES AND REEDS

The history of our nation's Spaceport at the Cape Canaveral Air Force Station and the Kennedy Space Center is relatively short, but historians know the area more intimately as **Cape Canaveral**. It is strange that the Spaceport of today would be so prominently known as a part of the "Port to the New World", because it was referred to so intensely in nearly all major maps from as far back as the 16th century.

*Cape Canaveral is one of the best known and most venerable locations in America. Its history can be traced through six centuries stretching back to one of the first sightings of the North American coast. The name describes what 16th century Spaniards saw; Cape Canaveral was literally **a canebrake**; or as the Spanish Historical Dictionary translates it, **a place overgrown with canes and reeds**. The name Cape Canaveral is preceded by few others in our nation's geography.*

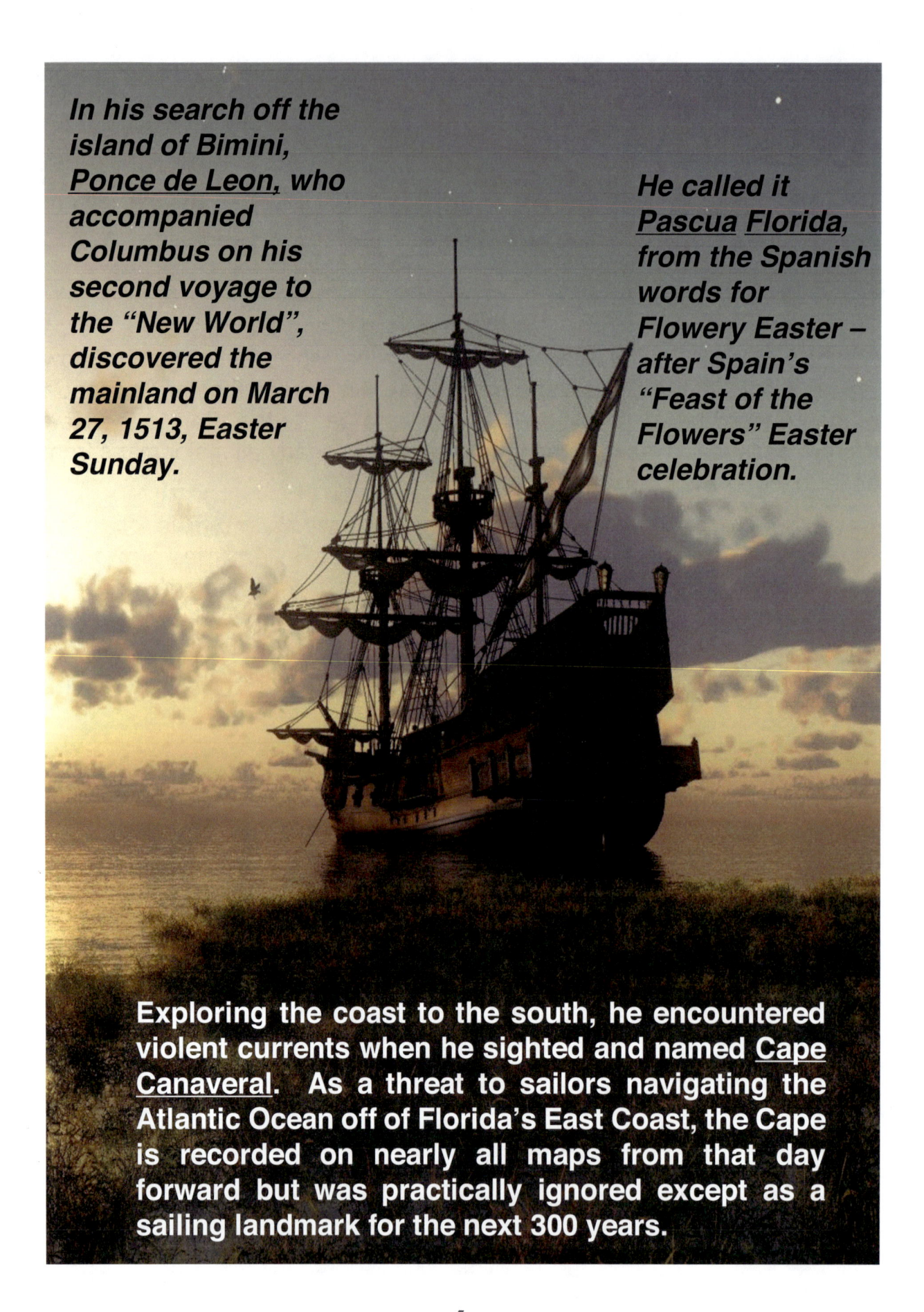

In his search off the island of Bimini, Ponce de Leon, who accompanied Columbus on his second voyage to the "New World", discovered the mainland on March 27, 1513, Easter Sunday.

He called it Pascua Florida, from the Spanish words for Flowery Easter – after Spain's "Feast of the Flowers" Easter celebration.

Exploring the coast to the south, he encountered violent currents when he sighted and named Cape Canaveral. As a threat to sailors navigating the Atlantic Ocean off of Florida's East Coast, the Cape is recorded on nearly all maps from that day forward but was practically ignored except as a sailing landmark for the next 300 years.

This same Cape was considered important enough by the founders of the United States that a lighthouse was established as one of the first 12 lighthouses required for our country.

Development came to stay in 1843 when the U. S. Government began construction of the Cape Canaveral lighthouse. It was 1853 before anyone came to live permanently however.

Captain Mills Olcott Burnham, his wife and two small children are credited with the distinction of tending the lighthouse through hardships hardly understood today.

Captain Burnham maintained the Cape Canaveral Lighthouse for 33 years. It stands today, serving as it has for over 170 years, the only concrete link between the past and the future.

This, then, was the Cape: a wild area with a lighthouse, the Keeper's Cottages, and some outbuildings.

HISTORY - THE CAPE AND EASTERN RANGE

Our nation's technology developed by leaps and bounds during World War II, and when the allied forces overrunning German-held territory captured the V-2 rockets, including the technical brains involved, the pace towards space exploration quickened. What began with the launch of captured V2 rockets at White Sands, NM in 1946 quickly moved east. That same year a government committee began work to select a site for a **Long Range Proving Ground**. After three years of study involving sites in the Aleutian Islands, California, and Florida, they selected the Banana River Naval Air Station because its climate was suitable for year round operation and the area was isolated and lightly populated. Initially the **Joint Long Range Proving Ground,** it was transferred to the Air Force and became the **Air Force Missile Test Center** in 1951.

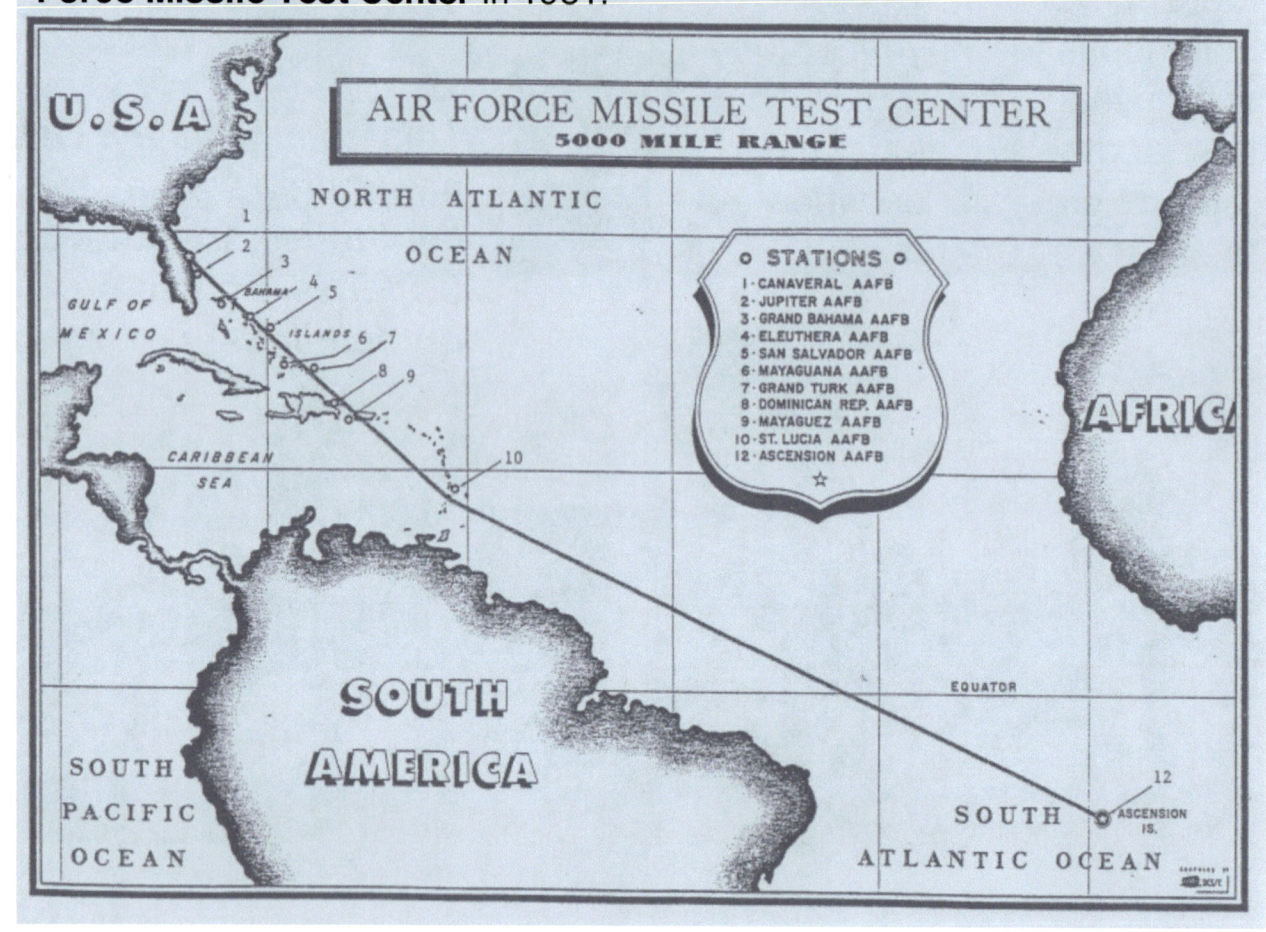

One of the key factors in that selection – perhaps the most significant – was the existence of 545 acres of land owned by the U. S. Coast Guard between the Cape Canaveral Lighthouse and the Atlantic Ocean – the Cape.

That area became a subdivision of the **Long Range Proving Ground** and by fall of 1949 work had begun on the Cape to develop launch pads 1, 2, 3, 4, and the island-based tracking stations in the Atlantic Ocean for downrange support. It became the **Cape Canaveral Air Force Station** in 1974 and in 2000.

FIRST LAUNCH

In May 1950, the first permanent construction work began at the Cape on a paved access road, and the first concrete launch pad (Pad # 3) was built. That pad became the launch site for the first launch from the Cape.

The "Blockhouse" for this firing was a wooden building, and the gantry was scaffolding on wheels built by the Army. The second and final Bumper launched successfully five days later from this same pad.

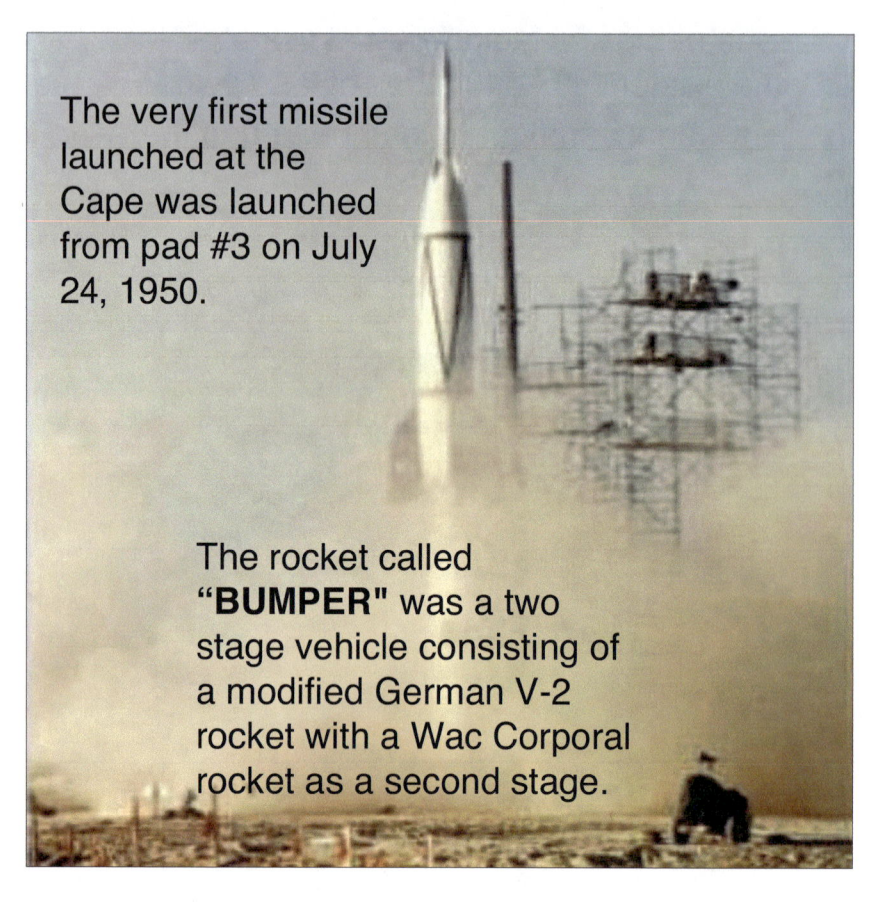

The very first missile launched at the Cape was launched from pad #3 on July 24, 1950.

The rocket called **"BUMPER"** was a two stage vehicle consisting of a modified German V-2 rocket with a Wac Corporal rocket as a second stage.

NEXT UP...

Three months later, October 25, 1950, the first **LARK** missile was launched successfully from Launch Pad #3.

The Lark, a winged, pilotless surface-to-air missile powered by a two-stage solid rocket system, flew a range of almost 10 miles. The Lark was never developed beyond the prototype phase.

FIRST DOWN-RANGE FLIGHT – *OFF WE GO!*

Eight months later, June 29, 1951, the first flight of the MATADOR missile was fired from a truck-mounted "zero-length launcher" and sand bag bunker east of pads 1 and 2 for launch control. Down-range instrumentation was used for the first time at Grand Bahamas Island, opening the range to 200 miles. MATADOR was a rocket-assisted jet aircraft, termed a "Pilotless Bomber".

On November 30, 1951, Construction was completed on the Cape's **CENTRAL CONTROL BUILDING.**

Central Control Building - 1952
(Photo: Air Force Space and Missile Center Gallery)

This facility provided a central point of control for all missiles and rockets launched from the Cape.

FACILITIES AND INFRASTRUCTURE

Carving a launch site out of the boondocks!

Remember that during this time, these gaudy red and white missiles were classified as "Top Secret", and the Cape was a "Hush-Hush" operation. Although new test and launch activities were ramping up daily, the Cape was very limited in its on-site capabilities, and heavy construction was underway.

A Missile **Skid Strip** was under construction at this time (1951). The Skid Strip was built to allow a guided missile with skids instead of landing gear to launch from the Cape, fly down range, and return to land on the skid strip on mission completion. It was five years before this took place, but the **Skid Strip** was, and still is, used as an airport for the Cape.

A BOMARC missile arrives at the **Skid Strip**. (Photo: - Air Force Space and Missile Museum Gallery).

A missile makes it way from Patrick to the Cape

The Cape was still an overgrown scrub land. Missiles were assembled and checked out at Patrick Air Force Base and transported by road over Route A1A through downtown Cocoa Beach to the Cape.

THE COLD WAR "SPACE RACE" IS ON...

As new facilities and launch sites came online, the pace of activities at the Cape accelerated rapidly. Range safety and missile control became key aspects of work at the cape, and the Central Control Building was completed in 1952. Initially the focus was on winged missiles and pilotless bombers.

On August 29, 1952, the first **SNARK** missile was launched from Launch Pad #4. The SNARK was an air - breathing missile.

On September 10, 1952, the first **BOMARC** was launched from Pad 4.

The BOMARC was a long range surface to air pilotless aircraft designed for high altitude flight to intercept hostile aircraft.

Central Control Building at the Cape
(Atlantic Missile Range - Air Force Missile Test Center)

On August 20, 1953, the first Ballistic Rocket made in the United States was launched from Pad #4. The **REDSTONE** was the largest missile launched from the Cape to this date. Later it launched from Pads 3-4, 5-6, and 26.

Developed by the U.S. Army and the Chrysler Corp, **REDSTONE** was a high-accuracy, liquid propelled, surface-to-surface missile capable of transporting nuclear or conventional warheads against targets at up to approximately 200 miles. It was 62 feet long and used alcohol and liquid oxygen as propellants.

During the 1950s the Cape was continuing to build up, and there were many "firsts" during this period of Cape history.

The NAVAHO was a two-stage, surface-to-surface, ramjet powered intercontinental missile, manufactured by the North American Aviation Corporation of Downey, California for the Air Force.

On August 19, 1955, the first launch of a **NAVAHO** was conducted at Launch Complex 9/10. The program was canceled in 1957 because the new **Intercontinental Ballistic Missiles (ICBMs)** were in development. Launch Complex 9/10 has since been disassembled.

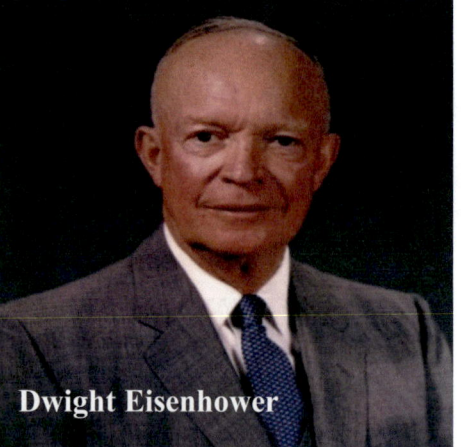

On September 8, 1955, President Eisenhower directed that the **ICBM** be given highest national priority, and the Cape took on a new and even more important role in the security and defense of the United States. The **ICBM** race with the Communist world – a key part of the "cold war" – was on.

Dwight Eisenhower

Launch Complexes **11, 12, 13, 14, 15, 16, 19, & 20**, known as **ICBM ROW**, were contracted for, with planned usage in 1957.

13

On March 14, 1956, the **JUPITER A** missile was tested at Launch Complex 26.

The JUPITER was an Intermediate Range Ballistic Missile **(IRBM)**, developed by the Army Ballistic Missile Agency **(ABMA)** in conjunction with Chrysler Corporation at Redstone Arsenal, Huntsville, Alabama.

The Jupiter Missile was designed for use against ground targets at ranges up to 1500 nautical miles. It is 60 feet long and approximately 9 feet in diameter. The missile is launched in a vertical position from a small steel platform. Propellants consist of cryogenic liquid oxygen and a kerosene-based fuel called RP-1.

On December 8, 1956, the first test vehicle of the **VANGUARD** program was successfully launched from Complex 18. Project VANGUARD was a NAVY project from the Naval Research Laboratory and was initially funded by the National Science Foundation as part of the 1957 International Geophysical Year.

U. S. NAVY VANGUARD

The VANGUARD was a 3-stage rocket, approximately 72 feet long and less than 4 feet in diameter. It was to carry the satellite to a 400-mile altitude and inject it into earth orbit at 18,000 miles per hour. The first stage was only 45 inches in diameter and its single chamber engine only produced 27,000 pounds of thrust. The vehicle weighed 25,000 pounds. The second stage was the first hypergolic-fueled rocket and the third stage contained a solid propellant rocket with a break-away nose cone shielding the satellite. VANGUARD was designed and built by the Glenn L. Martin Company of Baltimore, Maryland (now Lockheed Martin Corporation).

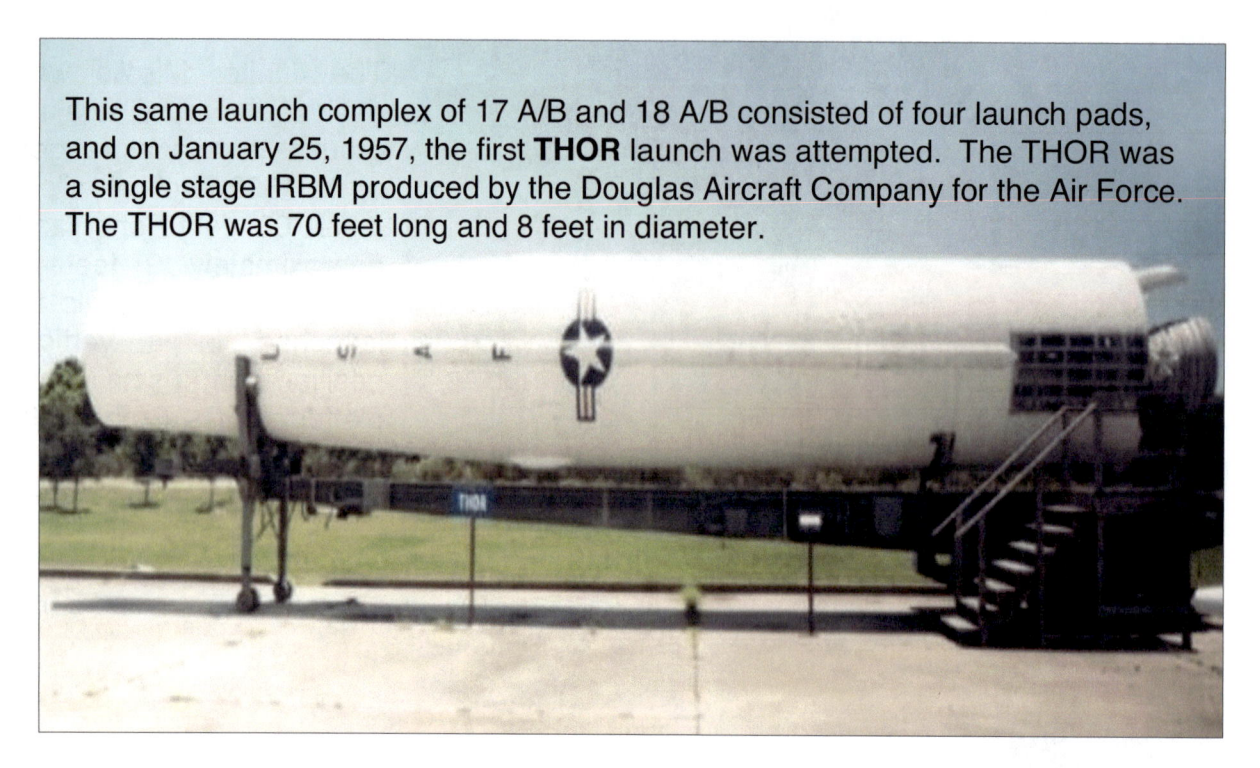

This same launch complex of 17 A/B and 18 A/B consisted of four launch pads, and on January 25, 1957, the first **THOR** launch was attempted. The THOR was a single stage IRBM produced by the Douglas Aircraft Company for the Air Force. The THOR was 70 feet long and 8 feet in diameter.

The year 1957 was one of mixed emotions for the U.S. The first THOR IRBM had a catastrophic first launch failure in January, while the Russian launch of **SPUTNIK** into earth orbit on October 4, 1957, served notice to all the peoples of the world that they, not the U.S., were ahead in the "Race-for-Space". This environment increased the U.S. efforts to both success and further failures.

On March 13, 1957, the **BULLGOOSE** missile was successfully launched from Complex 21/22. The BULLGOOSE was essentially another air breathing air-to-surface missile. It was a diversionary missile whose main mission was the protection of SAC Bombers. Fairchild Aircraft of Hagerstown, Maryland was the prime contractor.

BULLGOOSE was a 2 stage vehicle about 60 feet long with delta wings of a span of 31 feet and body diameter of 5 feet.

One month later, April 13, 1957, the first **POLARIS** was successfully launched down the Atlantic Missile Range from Complex 25/29. POLARIS was a single stage solid propellant rocket capable of being launched from a submerged nuclear powered submarine. It was an IRBM with a range of 1200 to 2500 nautical miles.

Named for the North Star, POLARIS was 28 feet long, 4-1/2 feet in diameter and weighed 28,000 pounds. The Prime Contractor was Lockheed Aircraft Company for the NAVY. The important thing about POLARIS is that, combined with its nuclear powered submarine launcher, it was the most versatile weapon this country had for that day.

The ICBM was of tremendous importance to the defense status of the U.S. 21 months after President Eisenhower directed top priority for ICBM development, the first **ATLAS** ICBM blasted from Complex 14 on June 11, 1957. The ATLAS was the first operational ICBM. It was capable of delivering nuclear payloads to a range in excess of 5,500 miles.

Prime contractor was General Dynamics Corp. of San Diego, California. The **ATLAS** airframe averaged 80 feet in length and 10 feet in diameter. The distinguishable feature of the ATLAS are the two lateral pods mounted on opposite sides of the tank structure, which was a light-weight stainless steel, thinner than a dime, pressurized to maintain its rigid shape. You might liken the structure to that of a thin steel balloon.

The ATLAS program development utilized four launch complexes along what is known as **"ICBM Row"**. In addition to Complex 11, there were also Launch Complexes 12, 13 and 14, continuously operational for the ATLAS program.

The **JUPITER C** missile was fast proving its capability, and on August 8, 1957, a nose cone from a **JUPITER-C** was recovered from the South Atlantic Ocean. This was the first recovery of a nose cone on the range, but the important thing about that flight was the letter carried in the nose cone, addressed to Major General John Medaris of the Army Ballistic Missile Agency - the first ever "**SPACE MAIL**".

On December 6, 1957, the first U. S. attempt to launch an earth satellite ended in disappointment when the Navy's **VANGUARD** rocket failed at launch.

In addition to being the nation's first satellite launch, this was the International Geophysical year -- and the VANGUARD was to be the Satellite Launching Vehicle.

This was the most publicized scientific failure in the U.S. space effort and another blow to our international ego.

1957 came to a close on a note of failure, but 1958 was a turning point for the U. S. space program and proved to be a "bumper" year for the Cape.

As shown in Table 1, there were more than 145 launches from the Cape that year, involving 15 different launch vehicles.

Table 1. Cape Launches in 1958*

LAUNCH VEHICLE	ATTEMPTS	PADS USED
ATLAS	14	11, 12, 13, 14
ALBM199C	2	B-58 Aircraft
BOLD ORION	8	B-47 Aircraft
BOMARC	20	4
BULL GOOSE	13	21, 22
JASON	6	10
JUNO	7	5, 26
JUPITER	6	6, 26
MATADOR	10	Zero Distance
NAVAHO	6	9, SKID STRIP
POLARIS	8	25
REDSTONE	7	5, 6
SNARK	14	1, 2
THOR	19	17, 18
VANGUARD	6	18

*Data by Clifford Lethbridge, Spaceline Inc., www.spaceline.org.

On January 31, 1958, **EXPLORER 1 was the first U.S. satellite** to be placed into earth orbit. The rocket was a **JUPITER-C** that was launched from complex 26 by the Army Ballistic Missile Agency **(ABMA)**. The **EXPLORER 1** satellite weighed 30.8 pounds. Mission duration was 111 days, and it remained in orbit until 1970.

EXPLORER 1 DISCOVERED THE VAN ALLEN RADIATION BELT THAT CIRCLES THE EARTH

William Pickering, James Van Allen, and Wernher von Braun holding Explorer 1. A model Jupiter C rocket is on the right.

On March 17, 1958, the second U.S. satellite, **VANGUARD 1**, was blasted into earth orbit from launch complex 18 B on a **NAVY Vanguard** Rocket. Payload capacity was 25 pounds.

The rocket with its satellite fairing was 72 feet long. It was a three stage rocket burning kerosene and liquid oxygen for stage 1, hypergolic chemicals for stage 2 (they ignite on contact), and a solid rocket for the stage 3.

THE VANGUARD 1 PAYLOAD WAS THE FIRST SUCCESSFUL SATELLITE OF THE VANGUARD SERIES; THE FIRST TO USE SOLAR CELL POWER. IT IS THE OLDEST SATELLITE STILL ORBITING THE EARTH.

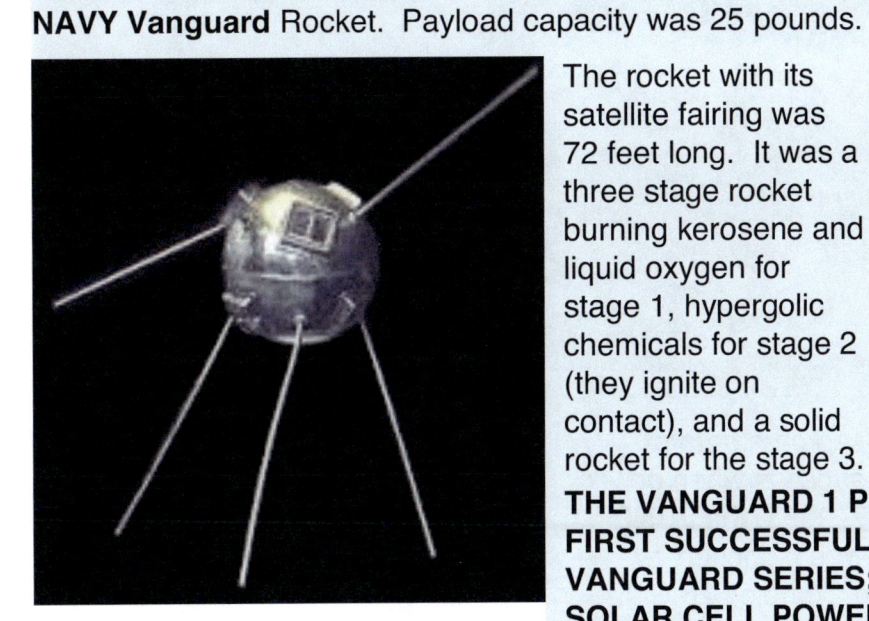

VANGUARD 1 SATELLITE

On April 3, 1958, the first launch of a THOR-ABLE was accomplished from complex 17. The THOR-ABLE was a modified THOR IRBM with an ABLE second stage. It had 15,000 pounds more thrust than its military counterpart.

On August 27 a **JASON** High Altitude Sounding Rocket launched from Pad 10. Jason was a five-stage solid rocket capable of lifting 125 pounds to 500 miles.

The development work during this time used a variety of methods and technologies for launch and recovery. "Air Drop" as well as horizontal and vertical launch methods were used at the Cape. The **BOLD ORION** and **ALBM199-C** rockets were ballistic missiles launched from aircraft – a **B-47** for the **BOLD ORION** and a **B-58** for the **ALBM199-C**.

ALBM199-C

1958 closed on a note of complete success when on December 7, 1958, the **JUNO II** missile was launched by ABMA from Complex 26 with the Pioneer 3 lunar probe.

BOLD ORION

19

"MANNED" SPACEFLIGHT BEGINS

On July 29, 1958 Congress and the President of the United States created the NATIONAL AERONAUTICS and SPACE ADMINISTRATION (NASA). The year 1959 would add new interest and complexity to the growing "Space Race" as men began to train to serve as flight crews for both Soviet and American spacecraft.

On April 2, 1959, the first seven ASTRONAUTS were selected by NASA for the **MERCURY** "Man-in-Space" project. They were known as the **MERCURY SEVEN**.

The MERCURY SEVEN in front of an F-106 Delta Dart

L to R: **SC0TT CARPENTER, GORDON COOPER, JOHN GLENN, GUS GRISSOM, WALLY SCHIRRA, ALAN SHEPARD, and DEKE SLAYTON.**

These first U.S. Astronauts are also known as the "Omega Seven" and are memorialized by the only monument on Cape Canaveral Air Force Station.

In addition to the "new" activity surrounding manned spaceflight, 1959 continued the frenzied activities at the Cape when the first Titan 1 ICBM was launched from Complex 15.

A Titan 1 launch from the Cape on the Atlantic Missile Range.

The **TITAN I** was the United States' second generation ICBM, and was developed by the Martin-Marietta Corp. in Denver, Colorado.

The **TITAN I** was a two stage vehicle over 93 feet long. Total weight at lift-off was 110 tons.

This rocket could carry a 15-megaton yield nuclear warhead to 5,500 nautical miles with speed in excess of 17,000 mph.

To give you some idea of what this speed means, if you were to travel from the Cape to Miami, it would take all of 45 seconds!

The TITAN I and TITAN II used complexes 15, 16, 19, and 20 along "ICBM Row"..

On October 29, 1959, the first **MACE** was rocketed from Complex 21/22. This complex was a "hardened" prototype launch site capable of withstanding nuclear attack. The **MACE** was an advanced version of the famous **MATADOR** missile. It was an Air Force surface-to-surface missile produced by the Martin Company of Baltimore, Maryland. **MACE** was 44 feet long, 4 1/2 feet in diameter, with a wing span of 23 ft.

THE CAPE CONTINUES TO BUILD UP...

The missile assembly location had been moved to the new Cape INDUSTRIAL AREA. This new complex spread to the west and south of the CENTRAL CONTROL BUILDING, and consisted of new hangars (missile assembly areas), a hospital, cafeteria, fire house and many other buildings. A sign from 1959 provides a good idea of those involved at that time. (Photos from the Air Force Space and Missile Center).

The year 1960 continued this country's success in space when the first launch of the newly developed **PERSHING** left Complex 30 on February 25, 1960.

The **PERSHING** was a two-stage solid propellant rocket, designed and developed for the ARMY by the Martin-Marietta Corp. at Orlando, Florida.

PERSHING was a Tactical Range Ballistic Missile (TRBM) capable of carrying a nuclear or conventional warhead. It was the replacement for the **REDSTONE** and was on constant duty in Europe. This missile is 34 feet long, 3 1/2 feet in diameter and weighs 10,000 pounds. **PERSHING** used the ARMY concept of assembly on site and represented the modern artillery piece.

MERCURY – The FIRST SPACE CAPSULE

1960 was a year of extreme activity at the Cape, with initial activities leading to the landing on the Moon:

- April 1, 1960 witnessed the launching of the World's first **WEATHER SATELLITE, TIROS I**. **THOR-ABLE** was the Satellite Launch Vehicle.
- On May 20, 1960, an **ATLAS ICBM** flew 9,000 miles to the Indian Ocean.
- On July 20, 1960, the first **POLARIS** launch from a submerged submarine took place from the nuclear submarine USS George Washington.
- On August 12, 1960, **ECHO 1**, the 100 ft. inflatable sphere satellite was placed into earth orbit by a **THOR-DELTA SLV**.
- September 21, 1960, the first Air Force **BLUE SCOUT** was fired from Complex 18.

The year 1960 closed on a note of complete success when the first **MERCURY-REDSTONE** launch on December 19, 1960 lobbed the first MERCURY Capsule on a ballistic path 205 nautical miles down range.

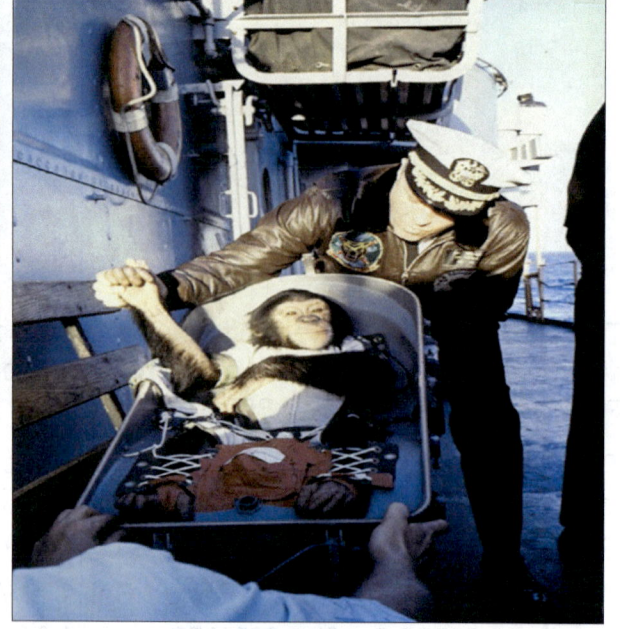

The American "march-into-space" continued when on January 31, 1961, a second MERCURY capsule was launched by a REDSTONE to carry a 37 pound Chimpanzee named "HAM" on a ballistic path 418 NM down range. HAM completed that first flight in great shape, demonstrating the capability of primates to survive the rigors of the launch.

On February 1st, 1961, the first **MINUTEMAN** was launched from Complex 32. This was more than just another "first", as this was the first time all stages of a multi-stage weapon system were tested on the initial attempt. **MINUTEMAN** was also the first all solid propellant ICBM.

MINUTEMAN was developed by the Boeing Airplane Company of Seattle, Washington for the Air Force. It was a three-stage solid propellant rocket with a range of 6,300 statute miles. It carried a nuclear warhead at speeds over 15,000 mph. The airframe is 57 feet long with a maximum diameter of 6.2 feet at a weight of 65,000 pounds. **MINUTEMAN** had extremely long term storage capability and was launched from underground silo-type facilities.

Just as the Sputnik disillusionment was dimming, Yuri Gagarin, a Russian Cosmonaut, was placed into earth orbit on April 12, 1961, and our national pride took another worldwide "back seat".

However, like the tortoise and the hare story, one month later on May 5, 1961, the first manned **MERCURY-REDSTONE** with the spacecraft **FREEDOM 7** flew a 260 NM sub-orbital ballistic flight carrying **ALAN B. SHEPARD, JR.**

Although this flight did not achieve orbit, it was the first U. S. manned flight, opening a new chapter in the U. S. space program. Remember this man –
**Alan B. SHEPARD, JR. –
First U.S. man-in-space.**

AMERICA'S "RACE FOR SPACE"

The "Race-for Space" was definitely on, and on October 27, 1961, the first **SATURN I** was successfully launched from Complex 34.

SATURN I was a two-stage clustered engine booster. The first stage contained 8 engines of 165,000 pounds of thrust each for a total of 1.32 million pounds of thrust.

Stage 1 was built by Chrysler Corporation for NASA. Stage 2 was designated S-IV and was built by Douglas Aircraft at Santa Monica, California. The second stage was later powered, but the first flight was loaded with water ballast and the spacecraft was simulated by dummy upper stages.

Saturn 1 Launch from Complex 34, October 27, 1961

The complete first **SATURN** I was 116 feet long and weighed almost a million pounds. At that time, Launch Complex 34 was the largest known rocket launch site in the world and the first to be built expressly for the peaceful exploration of space.

The following year, on February 20, 1962, the first U. S. manned earth orbital flight took place from the **ATLAS** pad 14.

LT. COL. JOHN H. GLENN JR. orbited the earth three times in the ATLAS launched MERCURY capsule **FRIENDSHIP 7**.

Remember his name –
JOHN H. GLENN, JR. –
First U.S. Astronaut to orbit the earth.

AND THE PACE QUICKENS!

On May 8, 1962, the first **ATLAS-CENTAUR** launched from complex 36. The **CENTAUR** was the Free World's first high-energy space vehicle.

The **CENTAUR's** first stage was a standard **ATLAS**, similar to those used for the Mercury project. The vehicle, less payload, weighs 293,000 pounds and is 133 feet high at launch. The new element of **CENTAUR** was the hydrogen-fueled twin engine second stage. **CENTAUR** was built by General Dynamics of San Diego, California for NASA.

President John F. Kennedy made three visits to Cape Canaveral during his term of office, and he alone is credited with the foresight to establish the "Man on the Moon" space goal for the United States.

President Kennedy's untimely death in November, 1963, resulted in **the renaming of NASA's Launch Operations Center to the John F. Kennedy Space Center**.

The Cape was uncommonly quiet at this time as very few launches were taking place. The ICBM Race was over! The Cape was losing its "hush-hush" reputation, and on December 15, 1963, the Cape was officially opened to the public on a regular basis.

THE CAPE BEGINS TO TRANSFORM

During this time (1963-64), construction was expanding to the north, and two great complexes were changing the unbroken "False Cape" area to one of major change. No longer could launch pads be close together, nor could the newer vehicles be assembled far away from the launch pads.

The result was the new "integrated" launch complexes. The prototype of these launch complexes was **Launch Complex 37.** It was similar to launch Complex 34, but contained more of the equipment necessary for the sprawling type of truly **Integrated Transfer or Mobile Launch Complexes** being constructed.

Launch Complex 37 was christened on January 29, 1964, when the fifth **SATURN**, SA-5 with the second stage powered, was launched. It orbited an unmanned optical payload.

The **Block II SATURN I** used the same first stage as the original **SATURN I**, except the booster was rated at a total of 1.5 million pounds of thrust.

The test proved the flight capability of the SATURN I liquid hydrogen clustered engine upper stage and demonstrated the vehicle's capability to orbit a 10-ton payload.

Apollo Test Flight A-102 flew on SA-7 from Complex 37 on September 18, 1964.

GEMINI – THE TWO MAN CAPSULE

GEMINI was the next step toward a manned lunar landing.
It bridged the flight experience between the short duration **MERCURY** missions and the long missions of **APOLLO**. The two-man **GEMINI** space craft was bell shaped, twice as heavy, 20% larger, and contained 50% more cabin space than the **MERCURY** space craft. The first launch of **GEMINI** took place on April 8, 1964, from Complex 19.

The **GEMINI** first manned flight, **GEMINI III**, was launched from Complex 19 on March 23, 1965. The Astronauts were **GUS GRISSOM and JOHN YOUNG**.

The **GEMINI** program consisted of 12 successful flights of which 10 were manned. **GEMINI** provided the first American demonstration of orbital rendezvous -- a skill which had to be developed to land U.S. Astronauts on the moon.

The launch vehicle was the modified **TITAN II** rocket. The target vehicle for the **GEMINI** rendezvous and docking missions was a modified **AGENA-D** vehicle with a forward mounted target docking adapter. The **AGENA-D** was launched on a modified **ATLAS ICBM** from Complex 14.

Major objectives achieved during the **GEMINI** program included on orbit maneuvering, orbital rendezvous and docking, extra-vehicular operations, and demonstration that man can perform effectively during extended periods in space both within and outside the spacecraft as key elements for Apollo.

THE AIR FORCE "TITAN" PROGRAM

While NASA got most of the publicity, the Air Force was very active!

On September 1, 1964, the first **TITAN IIIA**, a "Standard Core" for a Standard Space Launch Vehicle, was launched from the converted Complex 20.

This **TITAN IIIA** was a **TITAN II** with a third stage called a "Tran-stage". The complete **TITAN IIIA** was 10 feet in diameter and 124 feet long.

The **TITAN** family of ICBMs was still growing. The Integrate-Transfer-Launch (ITL) facility with pads 40 and 41 for the AIR FORCE **TITAN III** Standard Space Launch Vehicles were being constructed on the north end of the Cape.

The **TITAN III** ITL facility consisted of a **VEHICLE INTEGRATION BUILDING (VIB)**, a **SOLID MOTOR ASSEMBLY BUILDING (SMAB)**, **Launch Pads 40, 41**, connecting parallel railroad tracks, railroad engines and **TRANSPORTERS**.

The TITAN III **Vehicle Integration Building (VIB)** consisted of high bay assembly areas, supporting shops, labs, and the Launch Control Center (LCC). The VIB was located on a man-made island in the Banana River. The TITAN III Standard Core was assembled and checked out in the VIB, mounted on a Transporter, and moved over a parallel set of standard-gauge railroad tracks by two diesel engines to the SMAB.

The **SMAB**, Solid Motor Assembly Building, was located on another man-made island north of the VIB. At the SMAB the large solid rocket motors were assembled and checked out separately from the standard core.

When the Standard Core /Payload Assembly arrived at the SMAB, the Solid Motors were attached to the Standard Core while it is in the vertical position on the transporter. The **TITAN III-C** vehicle then continues to the launch pad on the **Transporter over the rail network**.

On June 18, 1965 the first **TITAN-III-C** launched from Launch Pad 40. The TITAN III-C was 124 ft. long with a 10 X 30 ft. oval area base and could place 25,100 pounds of payload into earth orbit. It can accelerate 6,400 pounds of payload into deep space or for lunar missions.

TITAN III-C #8 was launched on December 21, 1965, from Launch Pad 41, completing activation of this facility.

APOLLO – AMERICA'S MEN TO THE MOON

January 27, 1967 was the blackest day in the history of the U.S. Space effort. The APOLLO Spacecraft being readied and ground tested at **Launch Complex 34** caught fire and resulted in the deaths of **ED WHITE, GUS GRISSOM and ROGER CHAFFEE**.

This catastrophe nearly caused an end to this country's assault on the Moon. Only grim national determination and strong efforts by the entire APOLLO team prevented termination, and the APOLLO program survived.

The first three APOLLO launches before this black day were made using up-rated SATURN IB launch vehicles from Complexes 34 and 37. However, the delay caused by the fire during the fourth readiness effort resulted in a change in planned schedules.

The truly integrated mobile launch facilities of Launch Complex 39 were now ready for the giant SATURN V's.

This development spreads over the greater part of north Merritt Island and is designated as the **JOHN F. KENNEDY SPACE CENTER**. The facilities are located in two areas:

The **KSC INDUSTRIAL AREA** for administration and engineering personnel provided test facilities for the ground support equipment, the Apollo spacecraft, and astronaut training. A **VISITORS' COMPLEX** was located west of this industrial area and provided for public access to the Center.

LAUNCH COMPLEX 39: GATEWAY TO THE MOON

LAUNCH COMPLEX 39 included the **VEHICLE ASSEMBLY BUILDING (VAB)** where the rockets and spacecraft were assembled and tested, the **LAUNCH CONTROL CENTER (LCC)** housing the launch team for tests and fueling, the three mile long **CRAWLERWAY** over which the **TRANSPORTER** carried the **MOBILE LAUNCH PLATFORM and SPACE VEHICLE** to **LAUNCH PADS 39A AND 39B.**

The hub of operations at Launch Complex 39 is the **VAB**. Once the world's largest building by volume, the **VAB** is 525 ft. high, 716 ft. long, and 518 ft. wide. The building covers over eight acres. The **SATURN V** launch vehicle and spacecraft were assembled and tested in the VAB and mounted on the **MOBILE LAUNCHER**.

The **MOBILE LAUNCHER**, consisting of a **Mobile Launch Platform** Base and **Umbilical Tower** holding the giant **APOLLO/SATURN V**, was transferred from the VAB over a special Crawlerway by the **CRAWLER TRANSPORTER** to the Launch Pad 3-1/2 miles away.

The **CRAWLER TRANSPORTER** is 131 ft. long, 114 ft. wide, and moves on four double tracked crawlers. Each shoe on the crawler tracks weighs one ton. The **CRAWLER TRANSPORTER** weighs 5 1/2 million pounds and travels at a speed of one mile per hour.

The first APOLLO/SATURN V launched November 9, 1967 from
Launch Pad 39A. The full-up **APOLLO 4** was placed into unmanned earth orbit and initiated use of this massive Mobile Transfer Launch Facility.

All primary mission objectives were successfully accomplished and the U.S. space effort was on schedule.

The **APOLLO 5 & 6** flights further qualified the SATURN V space vehicle. On October 11, 1968, **APOLLO 7** launched on a Saturn1B from Pad 34 as the first manned APOLLO earth orbital flight, man-rating the COMMAND and SERVICE MODULES. The Astronauts were **WALLY SCHIRRA, DON EISELE and WALT CUNNINGHAM**.

On December 21, 1968, **APOLLO 8** was launched from Pad 39A. This was the first manned SATURN V flight to leave the earth, fly the first manned Lunar Orbit mission, and return to Earth. The astronauts were **FRANK BORMAN, JIM LOVELL and BILL ANDERS**.

The assault on the moon quickened, and on March 9, 1969, **APOLLO 9** was launched from Pad 39A. This flight fully qualified the LUNAR MODULE and set the stage for a lunar flight. The astronauts were **JAMES McDIVITT, DAVID SCOTT, and RUSTY SCHWEICKART**.

Two months later, on May 18, 1969 a Saturn V with Apollo 10 made the first launch from Pad 39B at KSC.

APOLLO 10 with astronauts **GENE CERNAN, JOHN YOUNG, and TOM STAFFORD** conducted the first manned low-level lunar orbital flight and rehearsed the efforts required to place a man on the lunar surface. **NASA was ready.**

APOLLO 11 – FIRST MEN ON THE MOON!

Less than two months later, APOLLO 11 launched from Pad 39A on July 16, 1969, and the U.S. placed a man on the moon on July 20, 1969.

Astronauts **NEIL ARMSTRONG, MICHAEL COLLINS and BUZZ ALDRIN** are credited with this climactic feat. This manned Lunar Landing achieved the national goal set by President Kennedy in 1961.

Apollo 11 – Flag on the moon

The Lunar Module, EAGLE, landed on the surface of the moon. NEIL ARMSTRONG, followed by BUZZ ALDRIN, photographed the moon's terrain, collected rock and soil samples, and transmitted TV pictures to over 25 million viewers in the U.S. and around the world.

They left a plaque that reads: **"We came in peace for all mankind".**

On November 14, 1969, the lunar landing mission was repeated by **APOLLO 12**. Astronauts **PETE CONRAD and ALAN BEAN** deployed experiments on the moon's surface while **RICHARD GORDON** piloted the Command Module for the mission. Parts of the unmanned **Surveyor III** that landed on the moon in 1967 were returned to earth.

Apollo 12 and Surveyor III on the lunar surface.

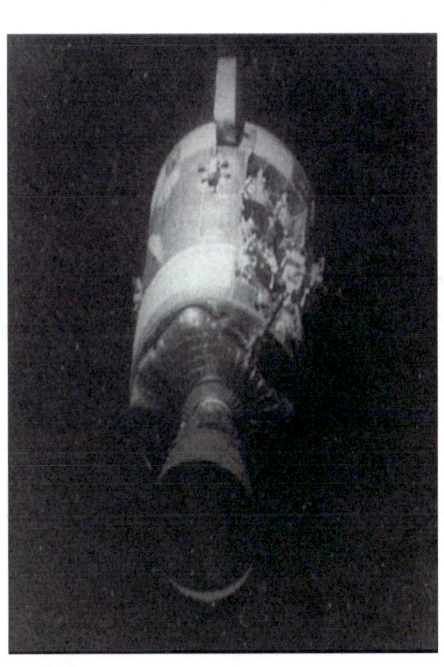

APOLLO 13 launched on April 11, 1970, from Pad 39A.

The Astronauts were **JIM LOVELL, JACK SWIGERT and FRED HAISE**.

The Service Module's oxygen tank #2 exploded two days after launch while on route to the moon, and the astronauts used the Lunar Module as a "life boat" to obtain a free-return trajectory around the moon, turning the Command Module back to earth orbit where the Astronauts were safely returned to splash down.

The world spent four suspenseful days of prayer for the safe return of these three brave men, and the U.S. space effort took another delay.

On January 31, 1971 **APOLLO 14** launched with **Alan Shepard, Stuart Roosa, and Edgar Mitchell** to explore the same area to have been visited by Apollo 13. It carried a two-wheeled Modular Equipment Transporter (MET) to carry lunar material.

APOLLO 15 launched on July 26, 1971 with **David Scott, Al Worden, and James Irwin** to survey and sample lunar surface features. It carried the first Lunar Roving Vehicle (LRV) allowing astronauts to travel farther from the Lunar Module.

APOLLO 16 launched on April 16, 1972. The crew was **John Young, Tom Mattingly, and Charlie Duke** and included an LRV for exploration.

The final lunar landing mission was flown by **APOLLO 17 on December 17, 1972**. The astronauts were **Gene Cernan, Ron Evans, and Harrison Schmitt** - the only trained geologist to walk on the moon. Their mission included extensive surface investigations using the LRV. **Gene Cernan** was the last astronaut on the lunar surface, leaving a plaque that reads: *"Here Man completed his first exploration of the Moon, December 1972 A.D. May the spirit of peace in which we came be reflected in the lives of all mankind."*

SKYLAB - THE FIRST SPACE STATION

The APOLLO Program consisted of six successful moon landings and demonstrated all of the requirements needed to develop future exploration programs. After Apollo, NASA launched **SKYLAB**, a program that created an early orbital workshop using the **S-IVB** upper stage from Apollo to serve as **the world's first "Space Station"**. Four missions were flown starting in May 1973. A **Saturn V** placed the S-IVB **Skylab** into low Earth orbit. There were three manned visits launched using a **Saturn 1B**. Missions ranged in length from 28 to 84 days, ending in February, 1974. These were followed by joint missions with the Soviet Union: **Apollo-Soyuz Test Project** and the **Russian Mir** space station.

SPACE SHUTTLE – *OUR SPACE TRUCK!*

The **SPACE SHUTTLE** Program was given Congressional approval during the U.S. Bicentennial in 1976, and work began to modify the Kennedy Space Center.

ORBITER PROCESSING FACILITY BAY 1

The **SPACE SHUTTLE ORBITER** required a special hangar, called the **ORBITER PROCESSING FACILITY (OPF),** and a new **SHUTTLE LANDING FACILITY (SLF)** with special landing guidance systems.

The SPACE SHUTTLE followed the same method of pre-launch processing as **APOLLO/SATURN V**. The complete STS was assembled in the **VAB** on the **MOBILE LAUNCH PLATFORM (MLP)** and moved from the VAB to the Launch Pad by the **CRAWLER TRANSPORTER (CT)** over a three-mile long **CRAWLERWAY**.

At the Pad the **SPACE SHUTTLE** went through a complete checkout for launch.

When the **Crawler Transporter** carrying the **STS** reached the Pad, the **MOBILE LAUNCH PLATFORM** was transferred to the Launch Pad and the CT was removed from the area. The Mobile Launch Platform was connected to the PAD facilities, including the **FIXED SERVICE STRUCTURE (FSS) and the ROTATING SERVICE STRUCTURE (RSS).**

The **SPACE SHUTTLE** used two four-segment **SOLID ROCKET BOOSTERS** (SRBs) and all three **SHUTTLE ORBITER** main engines firing for two minutes, after which the SRBs were parachuted to the Atlantic Ocean for recovery. The **ORBITER** and **EXTERNAL TANK (ET)** continued firing until orbital velocity was obtained, and the **ET** was then jettisoned to burn up in the atmosphere.

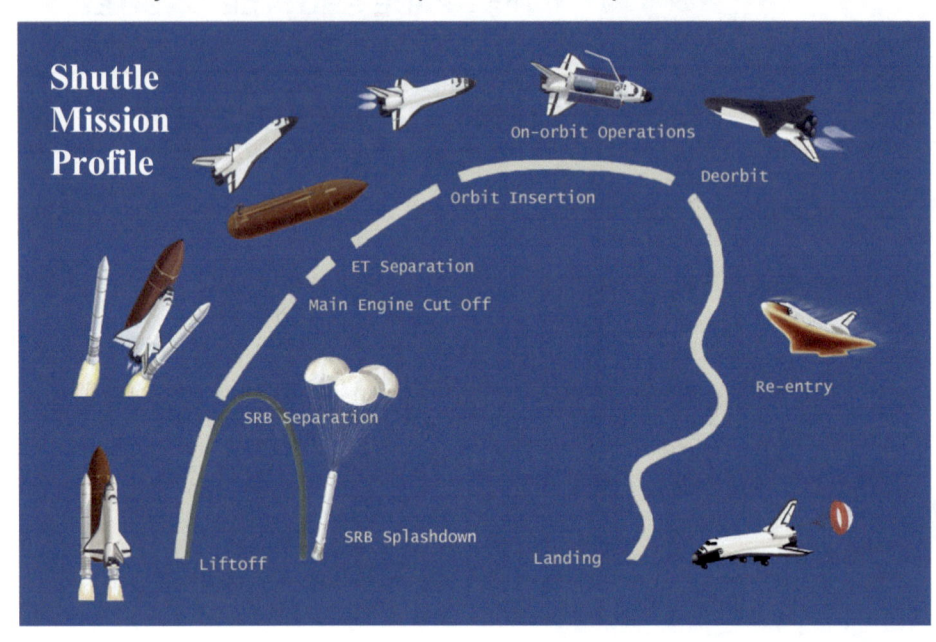

Shuttle Mission Profile

- On-orbit Operations
- Deorbit
- Orbit Insertion
- ET Separation
- Main Engine Cut Off
- Re-entry
- SRB Separation
- SRB Splashdown
- Liftoff
- Landing

The **ORBITER** entered earth orbit to conduct space experiments and a variety of missions.

At the completion of each mission, the **ORBITER** returned to earth as a lifting body, gliding unpowered for landing on an aircraft runway. The first flight landed on the dry lakebed at Edwards AFB, CA.

An important feature of the **SPACE SHUTTLE** was its re-usability. The **ORBITER** and **SRBs** were recovered and returned to KSC, either by landing at the **SHUTTLE LANDING FACILITY (SLF)**, transport via a **Boeing 747 (ORBITER)** or by rail cars for the **SRBs**.

The **ORBITER** was prepared for another launch and the **SRBs** were shipped to Utah, refurbished and returned to KSC. **SRBs** and **ORBITERS** were re-flown many times. The **SPACE SHUTTLE STS** made Cape Canaveral and KSC the Spaceport of Tomorrow.

SPACE SHUTTLE was a reusable **SPACE TRANSPORTATION SYSTEM** – the world's first. It flew a total of 135 missions over a 35-year span from April 1981 to July, 2011.

The first flight of the **SPACE SHUTTLE (STS-1)** using the **ORBITER COLUMBIA** was launched from Launch Pad 39A on April 12, 1981. Astronauts for this first flight were **JOHN YOUNG** and **BOB CRIPPEN**. The final flight was on the **ORBITER ATLANTIS**.

Some of the most memorable missions were: the first American woman in space (**Sally Ride**, STS-7); the first African-American in Space (**Guion Bluford**; STS-8); the first Spacelab placed into orbit (STS-9); the first in-space satellite repair (**Solar Max, STS-41C**); launching the **Hubble Space Telescope** into orbit (STS-31); **repairing Hubble** (STS-61); launching America's oldest astronaut into orbit (77 year-old Mercury astronaut **John Glenn**, STS-95); and the mission to begin building the **International Space Station** (STS-88).

TWO SHUTTLE CREWS FOREVER LOST

During its 30 years of flight the Shuttle Program experienced two major failures resulting in the loss of its flight crews. After twenty-four successful flights made from Launch Pad 39A, **the first launch from Pad 39B on January 28, 1986**, ended in catastrophe.

CHALLENGER exploded after 73 seconds of flight. All seven crew members perished, (Front row) **MICHAEL SMITH. DICK SCOBEE, RONALD McNAIR.**

(Back row) **ELLISON ONIZUKA, CHRISTA MCAULIFFE, GREGORY JARVIS, JUDITH RESNIK**.

The SPACE SHUTTLE program went through a 32-month review and design updating before the flight program was resumed in September of 1988 with the flight of STS-26 (ORBITER **DISCOVERY**).

38

When flights resumed, the SPACE SHUTTLE program settled into a period of International Space activities. Astronauts from many foreign countries were trained and flew on the SPACE TRANSPORTATION SYSTEM, bringing with them a wide variety of experiments and equipment developed and manufactured in their home countries.

The Shuttle Program was moving forward with renewed confidence and a full complement of missions when, on **February 1, 2003, the STS-107 COLUMBIA** Orbiter broke up as it returned to Earth over the west coast of the United States, killing all seven crew members. **NASA suspended Space Shuttle flight operations** for more than two years as it investigated the disaster, and major changes were made to eliminate damage to the ORBITER from debris during launch.

(L to R) **DAVID BROWN, RICK HUSBAND, LAUREL CLARK, KALPANA CHAWLA, MICHAEL ANDERSON, WILLIAM McCOOL, ILAN RAMON.**

During the SPACE SHUTTLE program delay, construction of the ISS was put on hold, and the Russian Space Program was able to provide added support for maintenance of the INTERNATIONAL SPACE STATION. **Shuttle operations resumed with return to flight on STS-114, ORBITER Discovery, on July 26, 2005**, 29 months after the loss of Columbia. Post-flight analysis showed debris falling from the external tank, and additional modifications delayed additional flights for a year, when **STS-121** launched on a **Return to Flight** mission to the ISS on **July 4, 2006**.

THE SHUTTLE ERA COMES TO AN END...

The INTERNATIONAL SPACE STATION (ISS) was placed into low earth orbit beginning in the year 2000 by a combined effort of the Space Shuttle Program and the Russian Proton boosters and Soyuz spacecraft. Sixteen countries have become active participants in the ISS, making it one of the most successful cooperative endeavors ever. Joint ventures and cooperative efforts have resulted in a strong co-existence among former competitors.

The Space Shuttle Program flew 135 missions and accomplished a wide variety of missions including primary scientific investigations into the effects of microgravity, testing of tools and processes for living and working in space, placement and retrieval of satellites, long duration exposure experiments, and construction and servicing of the International Space Station (ISS).

The Space Shuttle Program continued to fly resupply and crew exchange missions to the ISS from 2006 until the program ended with a fully successful flight of the **ORBITER Atlantis on July 21, 2011**.

The four remaining Space Shuttle ORBITERS were decommissioned and assigned to museums across the country for display to the American public.

Enterprise, the first orbiter built, is at the Intrepid Sea, Air & Space Museum in New York.

Discovery is at the Udvar-Hazy Center in Virginia.

Endeavour is at the California Science Center in Los Angeles.

Atlantis is on display at the Kennedy Space Center Visitor Complex in Florida.

There was no follow-on program and the U. S. remains without a way to independently fly crew members to the ISS, but that's about to change.

THE AIR FORCE EVOLVED EXPENDABLES

The "**EVOLVED EXPENDABLE LAUNCH VEHICLE**" **(EELV)** Program was initiated in 1994 by the Air Force to lower launch costs and assure access to space for military operations. EELVs grew from the **TITAN IV** program, which launched from Pads 40 and 41 beginning in June 1989.

The first new EELV to launch was the **Lockheed Martin ATLAS V** from Pad 41 on August 21, 2002 with a commercial communications satellite name Hot Bird 6 as its payload. It uses a Russian-supplied RD-180 engine burning kerosene and liquid oxygen that was first used on an Atlas III in 2000. The RD-180 remains the primary engine today, but U. S. engines are under development as a replacement by 2020.

On November 20, 2002, the **Boeing Company DELTA IV** was launched from Complex 37 -- the pad used for NASA's Saturn IB in the '60s – to place a Eutelsat communications satellite into geosynchronous transfer orbit.

The **DELTA IV** uses a Rocketdyne RS-68 engine, the newest engine since the Space Shuttle main engine (SSME) from the 1970s. Using an advanced design, it burns liquid oxygen and liquid hydrogen, making it a completely cryogenic system.

Both the Atlas V and the Delta IV were designed to serve U. S. Air Force's EELV program as well as commercial space requirements for medium and heavy payloads. However, costs and global competition made it difficult to compete successfully, and in 2006 The Boeing Company and Lockheed Martin formed a joint venture called the **United Launch Alliance (ULA)** to increase reliable access to space and reduce costs.

Using the Atlas V and Delta IV launching from both the east and west coasts, ULA prepares and launches payloads for military and commercial uses ranging from communications and weather satellites to interplanetary missions. Currently more than 30 Delta IV missions and more than 60 Atlas V missions have been flown, and they continue to provide the primary access to space for unmanned space programs.

INTERNATIONAL SPACE STATION AND NASA'S CONSTELLATION PROGRAM

In **1998 NASA** began construction of the **International Space Station (ISS)** using the Space Shuttle program as the primary launch system. **Crews boarded the station in 2000** and have been active on the station continuously since that time.

The ISS evolved as an international cooperative enterprise with resupply by both the U. S. and Russia. When the **Shuttle Program terminated in 2011**, the U. S. lost its capability to independently ferry crews to its space station and **has not yet recovered that ability.**

In **2004 NASA** embarked on a major new initiative to design, build, test and operate a space exploration program called **Constellation** designed as a replacement for the Shuttle Program to service the ISS and return to the moon and eventually travel to Mars.

Constellation was a full development program to replace the Apollo lunar landing program. It included two booster rockets **(ARES 1 & ARES 5),** an **Orion spacecraft**, and an **Altair crew habitat** for use on the lunar surface.

The **ARES 1** rocket was designed to use Space Shuttle-type solid rocket booster hardware to maximize technology transfer at minimum cost.

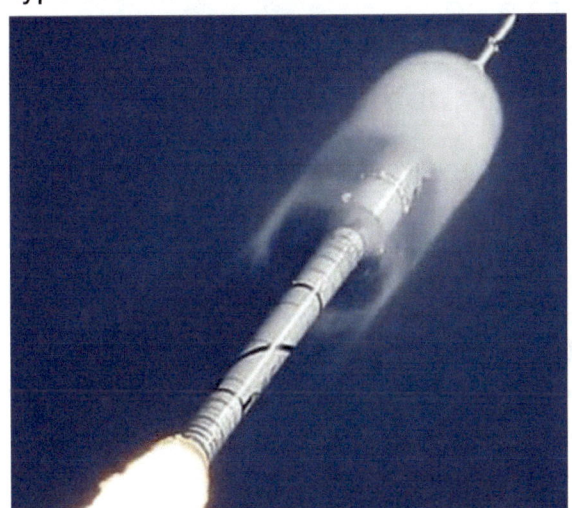

It flew a single test flight, **ARES-1X** from **Pad 39B** at Kennedy Space Center.

The program was funded for only **one test flight** before it was cancelled due to its high cost amid a major recession and a change in administration. **The space program faltered.**

42

THE CAPE CANAVERAL SPACEPORT TODAY

As the first decade of the 21st century drew to a close, **America's space programs were in major transition** across all three sectors: **military, civil, and commercial space**.

The Air Force was busy finding ways to reduce cost and expand pathways to space while moving forward with reusable flight hardware for orbital operations **(X-37B)**.

Air Force – Boeing X37B

NASA was hit with a major setback when their fledgling **Constellation Program** was cancelled. They continue to operate the **International Space Station** and a new **commercial crew program**.

NASA ARES-1-X Flight

Business and industry groups were eager to join a growing space sector that promised new opportunities with autonomy and profitable outcomes, and new partners began to emerge. These factors characterize and shape our space programs, and **a major transformation is now underway**.

Perhaps the biggest change is the **unprecedented cooperative effort among government agencies and private companies**. New initiatives are enhancing and expanding ongoing and future activities through shared use, re-purposing, renovation, and partnership activities among the participating organizations. **Utilization of facilities** across government agencies has been enhanced through activities by **Space Florida** – the state agency charged with business development of Florida's space related work – and a true **spaceport of the future** is emerging **at the Cape Canaveral Air Force Station** and the **John F. Kennedy Space Center**.

Air Force programs at the Cape are presently centered around Evolved Expendable Launch Vehicles (EELVs) prepared and launched by **United Launch Alliance** using the **Delta IV and Atlas V** launch vehicles.

These are characterized by the **Delta IV** launch on June 11, 2016 boosting a **National Reconnaissance Office (NRO) satellite** on a classified mission.

The **Delta IV Heavy** provides the greatest lift capabilities currently available for placing large/heavy payloads into orbit. The Heavy has lifted up to **57,156 pounds** (25,980 kg) into **Low Earth Orbit (LEO)** and up to **31,238** pounds (14,220 kg) into **Geosynchronous Transfer Orbit (GTO)**.

The **Delta IV** is primarily used for military payloads but is available for commercial and civil space activities including deep space missions when needed.

The **Atlas V** is also becoming the workhorse for military, civil, and commercial space launches. Recent flights have demonstrated the capability to lift up to **41,390** pounds (18,814 kg) to **LEO** and **19,580** pounds (8.900 kg) to **GTO**.

Recent launches for the **Atlas V** include a **MUOS-5** military **Comsat for the NAVY** aboard an Atlas V on June 23, 2016 and a **National Reconnaissance Office satellite** on July 28, 2016.

Both **Delta IV and Atlas V** have been used for military, civil, and commercial flights **including NASA deep space missions to Mars**. Much greater uses of both launch vehicles are already in the planning stages.

MOVING BEYOND EVOLVED EXPENDABLES

The newest addition to Air Force activities at the Cape is the preparation and launch of the **Boeing X-37B** orbiting space plane, also known as the **Orbital Test Vehicle**. This "mini-shuttle" is highly classified and launches from Complex 41 aboard an **Atlas V** rocket via ULA.

Nearing completion to house the **X-37B** are the modifications to use the **Shuttle Orbiter Processing Facility Bays 1 and 2** by the Air Force for preparation of the **Boeing X-37B**.

The **Shuttle Landing Facility** at KSC may be used for landing the **X-37B** after return from classified orbital missions.

As part of the strategy to strengthen its ability to access space quickly, reliably and at lower cost, the Department of Defense in the 1990s encouraged private companies to enter the space hardware development process. One of the companies accepting that challenge was **SpaceX**, which developed the single engine **Falcon 1** rocket using private funding in 2005/2006. SpaceX launched five Falcon 1 rockets from **Kwajalein Atoll** including a NASA payload on its second flight. The Falcon 1 was replaced in 2011 with the larger Falcon 9 rocket.

FALCON 1 at Kwajalein Atoll

In a major shift in the competitive posture of commercial space companies, **SpaceX Falcon 9** was recently certified for Department of Defense competitive bidding, broadening its base of operations to include **future Air Force contracts.**

FALCON 9 at KSC

THE NASA COTS and C3P PROGRAMS

Since 2011, the sole crew carrier to and from the ISS has been the **Russian Proton** rocket flying **Soyuz** capsules for rendezvous, docking, deorbit and landing in Kazakhstan. Two companies – **SpaceX (Dragon capsule)** and **Orbital Sciences/ATK (Cygnus Spacecraft)** – have been contracted by NASA as part of the **Commercial Orbital Transportation Service (COTS)** program to provide cargo carriers, and both have successfully launched to the Space Station for resupply.

The SpaceX **Falcon 9** rocket and **Dragon** spacecraft have become major contenders for NASA's cargo transport to the ISS. **Using 9 Merlin engines in the first stage** and a single Merlin engine in the second stage, development has accelerated due to military and commercial contracts for flights beyond NASA's human and cargo contracts, providing capital and demand for flights on a routine basis. The **Falcon 9** launches from **Complex 40**.

In addition to the **Falcon 9** rocket, the **SpaceX Dragon** was **the first private spacecraft to dock at the ISS.** Dragon has flown as an unpressurized cargo carrier and **will be upgraded to a human-rated capsule to replace the Russian Soyuz** as part of the **NASA Commercial Crew and Cargo Program (C3P).**

Orbital Sciences/ATK has successfully flown its **Cygnus** spacecraft on six missions and one failure to date as part of its **COTS** contract. An initial **ISS** docking mission and two **ISS** resupply missions were flown from **NASA's Wallops Flight Facility in Virginia** using an **Antares rocket.** Two **ISS** resupply missions were flown using an **Atlas V from the Cape.** A sixth resupply mission using **Antares** is in work.

The Orbital Sciences/ATK Antares Rocket and Cygnus Spacecraft provide another proven commercial cargo carrier under **NASA's (COTS)** program to fly resupply missions to the ISS. With **SpaceX and its Falcon 9 Rocket and Dragon Spacecraft, the American Space Program is close to having the capability to fly manned missions to the ISS aboard U. S. made rockets and spacecraft.**

NASA COMMERCIAL CREW SYSTEMS

NASA is working to replace the Space Shuttle using commercial crew systems now under development with three companies – **The Boeing Company** for the **CST-100 "Starliner"** capsule being fabricated at KSC; **SpaceX** and the **Dragon Spacecraft**; and the **Sierra Nevada Corporation** for its **"Dream Chaser"** lifting body space plane. Partnerships among military, civil, and commercial space entities are opening new avenues for exploring space at the Cape, KSC, and elsewhere.

The Boeing Company **Starliner** is being built in the former Shuttle Orbiter Processing Facility Bay 3 at KSC – a Space Shuttle hangar that has been repurposed for the NASA Commercial Crew Program. The building has been leased to Boeing, and renamed the **Commercial Crew and Cargo Processing Facility (C3PF).**

The **crew access arm** providing access to the **CST-100** capsule from the new crew access tower was installed at **Space Launch Complex 41 (SLC-41)** at the Cape on August 15, 2016. Significant changes in facilities and equipment are underway as part of NASA's Commercial Crew Program.

The ongoing cooperation among NASA, the Air Force, Boeing, and others demonstrates the extent of creative partnerships and joint activities now underway at the Cape and KSC.

The **CST-100** will fly aboard a **ULA Atlas V** from Complex 41 beginning in late 2017 or early 2018.

The **Sierra Nevada Corporation's Dream Chaser** spacecraft is a reusable orbital space plane designed to resupply the ISS with cargo and a crew of up to seven. An SNC photo showing the Dream Chaser linked to the ISS provides a good idea of how it will be used.

Dream Chaser is preparing for its Phase 2 flight testing from the **Armstrong Flight Research Center** in California, including free flying drop tests at that location. Vehicle fabrication is done in Louisville, Colorado.

DREAM CHASER AT ARMSTRONG FLIGHT RESEARCH CENTER, CA.

Launch of the **Dream Chaser** is currently planned from the CAPE using either an **Atlas V or a Falcon 9** – giving the **Dream Chaser** added flexibility for flight operations for a variety of uses. Landing will be to a horizontal runway in aircraft mode.

LOOKING BEYOND ISS TO NEXT STEPS

Following cancellation of the **Constellation** program, NASA was able to use the spacecraft design begun in that program for the **Orion capsule** – a larger version of the crew **Command Module** flown on **Apollo**. This is the spacecraft that can carry a crew to the moon and beyond.

Orion is funded as part of the **Commercial Crew Development Program** and has already flown three test flights on its way to becoming operational. In addition to its significance as a **major follow-on program** to Space Shuttle for crew, this commercial partnership program is bringing **manufacturing operations to the Kennedy Space Center** for the first time ever.

48

Working with Space Florida and NASA, Lockheed Martin was contracted and a former Apollo assembly area in the KSC **Neil Armstrong Operations and Checkout Building** (formerly the O&C Building used for Apollo) was converted to a fully operational manufacturing plant for the Orion Spacecraft.

This is a significant change in how business is done at KSC. Prior to Orion, KSC's role had been assembly, test, launch and recovery. **Orion brings the full spectrum** of aerospace development, manufacturing, test, launch, and recovery operations to KSC, providing the opportunity to expand deep space exploration beyond earth to the moon, Mars, and beyond.

Another change is use of operations that challenge past practices. For example, one of the key elements for success by **SpaceX** is its push toward **reusability**. The successful recovery of the first stage of its **Falcon 9** booster rocket and its **9 engines** will **reduce SpaceX reuse costs** significantly.

FALCON 9 Landing on a Drone-Ship at sea

FALCON 9 on Landing Site 1 at the Cape

In spite of strong doubts by some space veterans, **SpaceX has now recovered the Falcon 9 boosters** and all nine engines **six times** – **four** by water recovery using a **drone-ship**, and **two** by land using **Landing Site 1 at the Cape Canaveral Air Force Station.**

ULA is human rating the Complex 41 launch tower for commercial crew activities using the **Atlas V.** They will launch the **OSIRIS-REX** asteroid sample return mission this year, and they have won a NASA contract to launch the **Mars 2020** rover mission in the summer of 2020 as part of **NASA's Mars Exploration Program** to continue robotic exploration there, as well as NASA's **Interior Exploration Using Seismic Investigations, Geodesy, and Heat Transport (InSight)** mission in 2018.

MARS 2020 Rover

AMERICA'S FUTURE SPACE EXPLORATION

Although the nature of space exploration is fraught with changes in plans and cancelation of programs, changes in tenant policies and new partnerships have resulted in much greater use of current resources to spur space exploration in all areas. A series of current activities all point to continued growth and increasing uses of the Cape/spaceport for all manner of flights.

The NASA Space Launch System (SLS) will use a liquid-fueled core with 5 engines linked to two five-segment solid rocket boosters to provide up to 160 metric tons to low earth orbit. That level of lift will support missions to the moon, Mars and beyond.

The **VAB** is already being readied for this larger configuration, with completion of the platforms by December, 2016. Static firing of the engines and SRBs are already underway, and NASA continues to work toward an initial SLS flight in 2018. Nine moveable platforms to support the SLS build-up are in work, with more than half already installed and tested.

SpaceX has leased **Complex 39A for a hangar/assembly** area there, along with launch facilities at Complex 40. They have leased space at the cape for **SpaceX Landing Complex 1** and are working to add additional landing complexes to maximize reusability by returning all three stages of the Falcon Heavy rocket when it launches.

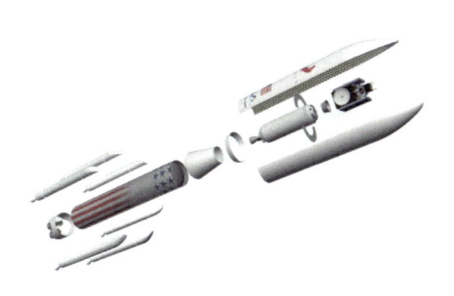

ULA is developing a **"next generation launch system"** (NGLS) rocket called **Vulcan** that will use a new American engine in place of the Russian RD-180 along with up to six solid rocket boosters and an Advanced **Cryogenic Evolved Stage (ACES)** in place of the Centaur upper stage used today to carry the largest payloads now forecast for future space missions.

NEW PLAYERS, PROGRAMS, PRACTICES

While **SpaceX, Boeing, Sierra Nevada, Lockheed Martin, ULA,** and others continue to work toward space station resupply, commercial crew support, and deep space missions, **new players continue to partner** at the Cape and at the Kennedy Space Center.

BLUE ORIGIN is leading a group of "newcomers" to the Cape for space activities. **Blue Origin** is a private company founded by Jeff Bezos, CEO of Amazon and headquartered in Kent, Washington. Over the past several years Blue Origin has built and flown reusable systems for suborbital flight from their **West Texas launch site**. Their **NEW SHEPARD** booster has flown numerous times, returning to the launch site in Texas to a soft landing.

In addition to designing and building their own rockets and spacecraft, **Blue Origin** has initiated work on new engine designs, including their **BE-4 ENGINE** now under contract for use by ULA as a replacement for the RD-180 Russian engine for **Atlas V**.

Blue Origin has **leased Complex 36 at the Cape** for launch and landing operations. Ground was broken in June 2016 for a very large **(475,000 sq. ft.) factory in NASA's Exploration Park at KSC** for operations to build and assemble components for the **New Shepard** system and **BE-3/BE-4 engines**.

Moon Express Inc., has licensed **Complexes 17 and 18 at Cape Canaveral Air Force Station** for its lunar lander development work in an effort to win the **Google XPrize** of $20M as the **first privately funded group to land a system on the moon** that can travel 500 meters and transmit video and images back to earth.

OneWeb – formerly WorldVU – has **leased land at the KSC Exploration Park** and broken ground **for a satellite manufacturing facility**. They plan to build as many as **660 satellites** and begin hiring in 2016. They will launch from multiple locations using the **Virgin Galactic Launcher 1** to **air launch from a 747** and a rocket carried to altitude. **In Florida** they'll use **SpaceX Falcon 9 and ULA's Atlas V**.

THIS IS JUST THE BEGINNING. THERE'S MUCH MORE TO COME!

YOU CAN JOIN IN EXPLORING SPACE!

Everyone can be a part of the space program at some level, and it can be easy and fun. Here are some "**secrets**" to start you **exploring space**. All the http:// links you see in blue are current. Type them into your browser to go to the sites.

How Rocket Engines Work...

Every kid (and MOST ADULTS) **enjoys watching rockets** "do their thing". Whether its fireworks, movies, or real launches into space, there seems to be some "**magic**" in how rockets work. However, once you understand how they work, *it's really pretty simple*.

Unlike cars that use wheels to push against the road, **boats or airplanes** that use propellers to pull themselves through water or air, **rockets generate power using "ACTION->REACTION" that makes them move** using only what they contain inside their tanks and engines. ***They need nothing from outside themselves***.

Like other "engines" **rockets burn fuel** – either liquids like kerosene with liquid oxygen, or solid fuels similar to matches. The energy produced by burning pushes against the sides and ends of a chamber that is open on one end, and the expanding gases provide the **ACTION** as they exit out the back of the rocket, forcing it forward as a **REACTION**.

To understand how rockets work, look at this picture and **think of what happens when you stand in the back of a boat and jump off**. As you go out the back (the stern), the boat moves forward toward the front (the bow). A skateboard will do that too. Try it and see for yourself. The bigger the person, the faster the boat or skateboard moves.

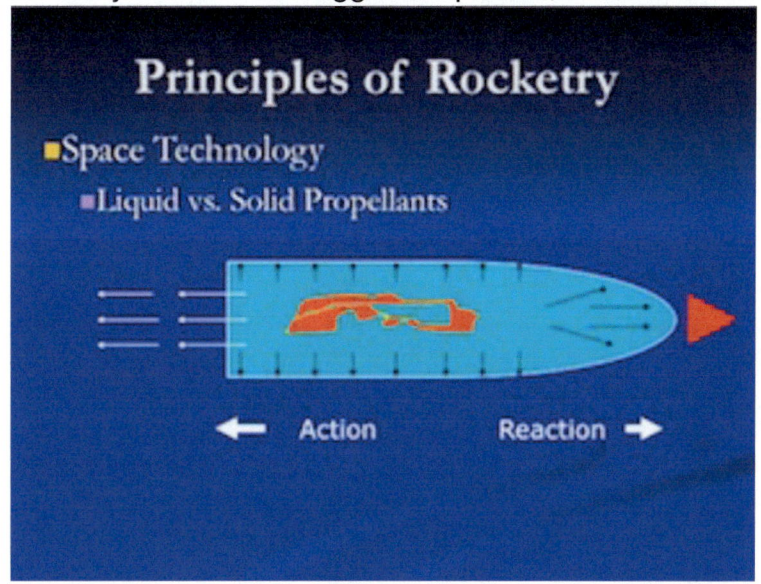

If you had an endless supply of sailors who could jump out the back of the boat, it would move forward faster and faster until you ran out of sailors. A rocket engine works just like that. Light it up and watch it go.

Unlike cars, boats, and airplanes, **rockets don't need anything outside themselves**. They carry everything they need to move forward until they run out of propellants. Cars, boats and airplanes need fuel too, but they also need oxygen from the air and friction from the road, water, or air to propel them forward. **Rockets need nothing from outside, because ACTION<->REACTION works on just what is inside the rocket.** That makes a rocket perfect for space, since space is mostly a vacuum (absence of air, water, etc.) and there is no friction to slow it down once it gets going. You can **learn more at**: http://science.howstuffworks.com/rocket1.html.

WHY IT MATTERS WHERE YOU LAUNCH

In the U. S., **launches take place from two main sites**: Florida via the Kennedy Space Center and the Cape Canaveral Air Force, and Vandenberg Air Force Base, California, based on **the type of orbit needed and the amount of energy available**.

On the **east coast, launches are to the east** to fly over water, avoid populated areas and take advantage of the 1,000 mph eastern movement of the Earth. Two main missions are **low earth orbit** to the International Space Station, and **geosynchronous orbit for satellites that need to stay over one spot** for communications and GPS. Geosynchronous flights match the rotational speed of the Earth to stay over one spot.

On the web coast, **launches are from the north to the south** to fly over water, avoid populated areas, and orbit satellites for observation, photography, and ground data by **passing over many areas of the Earth that rotates under the satellite each day**.

All orbits follow the same laws of physics – you must gain enough speed and height to **fall around the Earth for very long periods** of time without the need for more energy.

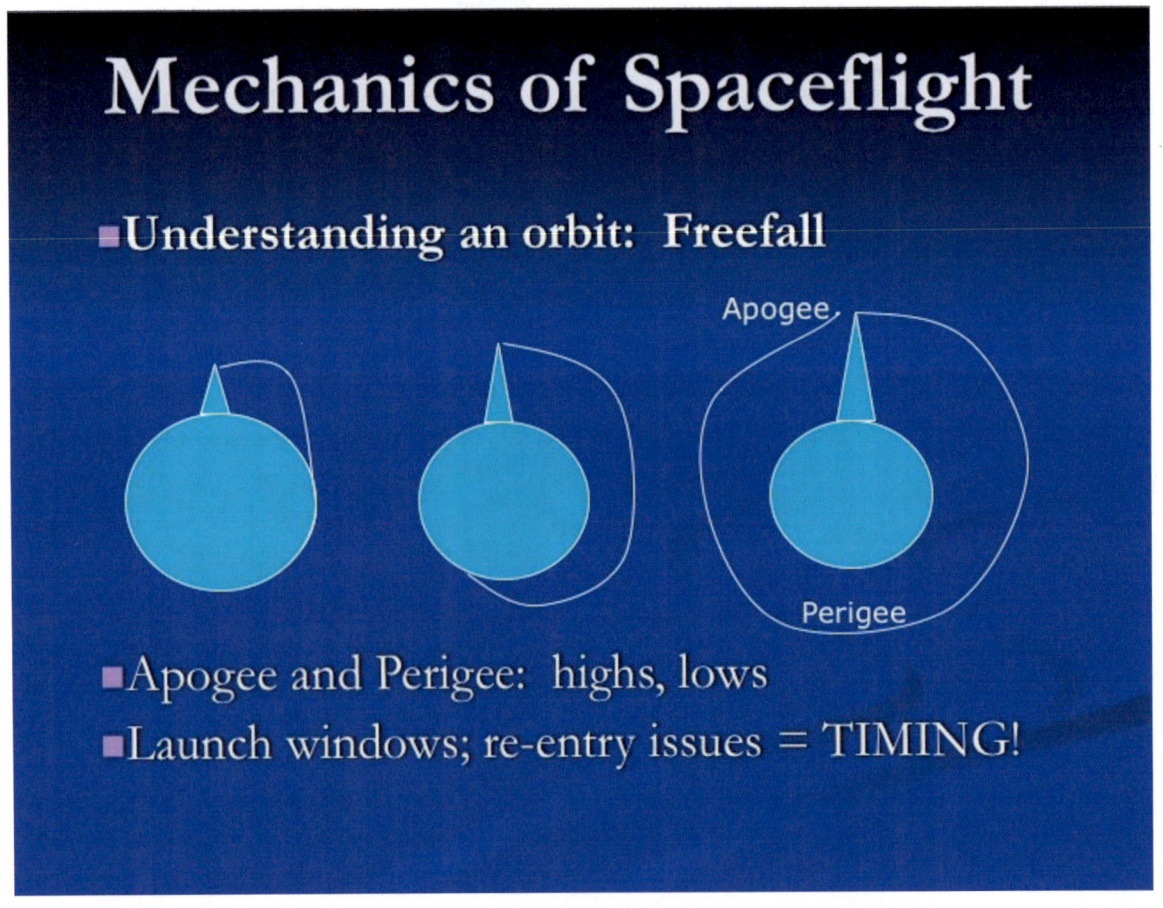

Mechanics of Spaceflight

- **Understanding an orbit: Freefall**

Apogee
Perigee

- **Apogee and Perigee: highs, lows**
- **Launch windows; re-entry issues = TIMING!**

Orbits are all free-fall processes, best explained this way. If you climb a tall mountain and shoot a cannon, the ball goes out and begins to fall to earth because of gravity. If you get a bigger cannon and use more powder, the ball goes farther, and if you go higher and get a big enough cannon, the ball goes out far enough that it continues to fall around the Earth at just the right speed and altitude so it never falls back to Earth. It's in orbit! For more information, see: http://octopus.gma.org/surfing/sats.html.

WHAT IT'S LIKE LIVING IN SPACE

There is a lot of interest today in exploring space and traveling to Mars much like the settlers who came to this country and traveled west as pioneers. When people go into space, they experience very different conditions than we have on Earth, and we are still learning how to live and work in space without causing permanent damage to ourselves.

One of the first things that happens when anyone goes into orbit is the body adjusts to micro-gravity. Remember, the spacecraft is in free-fall, like a roller coaster or elevator where you don't seem to weigh anything. You float, and your body fluids shift around. On Earth most of the liquid in our body is in our legs because of gravity on earth.

When we go into orbit, that liquid spreads throughout our body and we have to use the toilet to get rid of the excess fluids. **When we return to earth**, if we forget to drink water to get ready for gravity, **we'll faint as we step off the spacecraft** because gravity pulls fluids down to our legs and feet, and there's not enough left to get blood to the brain. That's just one example of the issues we are discovering as we learn to live and work in space. Some of the other concerns are listed in the following chart:

Living and Working in Space

- **Physiology of Weightlessness**
  - Fluid shift, muscle atrophy, bone loss
- **Maintaining Functionality**
  - Illness, Injuries, Interventions
- **Radiation Issues**
  - Cosmic rays – Short and Long Duration Flights
- **Returning to Earth**
  - G-loading, Remaining Conscious, Restoring Functions Lost in Flight

To learn more about the medical, biological, and physiological challenges we face as we explore space, see: **https://www.youtube.com/watch?v=SGP6Y0Pnhe4.**

HOW YOU CAN BE A PART OF ALL THIS!

There are lots of options for anyone who is interested in space exploration. One is **learning HOW TO FLY ROCKETS**, shown at: http://www.flyrockets.com/clubs.asp. For a look at **ALL the ways to become involved with NASA**, go to the NASA website at: http://www.jpl.nasa.gov/edu/learn/.

If you are able to travel, consider visiting one of the many museums and space-related sites where you can see first-hand some of the artifacts from our space history and participate in interactive programs designed to help you experience some of the key aspects of space-related jobs.

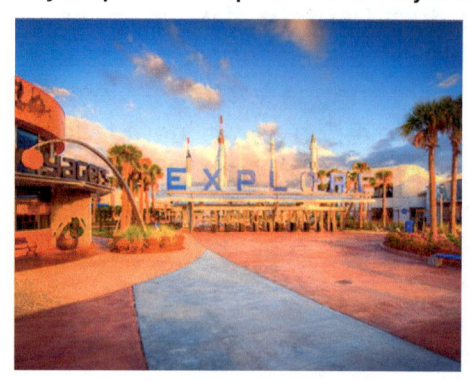

The Kennedy Space Center Visitor Complex in Florida provides the opportunity to "Meet an Astronaut", have lunch with an astronaut, see the Shuttle Atlantis, sit in the pilot's seat in a shuttle cockpit mockup, and tour the KSC launch pads and landing sites where rockets are prepared for launch and landing. A giant Saturn V rocket with real space hardware that was prepared to take men to the moon is there, and space veterans will explain it all. See https://www.kennedyspacecenter.com/.

The National Air and Space Museum in Washington D. C. provides access to a wide variety of aviation and space displays and artifacts, from a mockup of the Wright Brothers' airplane that achieved man's first flight to supersonic jets and real space hardware including a Skylab "space station" available for walk-through. See https://airandspace.si.edu/.

The U. S. Space Walk of Fame Museum in Titusville, Florida, offers a walk through the history of spaceflight with historic space suits and actual launch consoles used at the Kennedy Space Center and the Cape Canaveral Air Force Station. See http://spacewalkoffame.org/

The History Center of the Air Force Space and Missile Museum is free. See http://afspacemuseum.org/

If you know someone who worked in the space program, get their name into the Space Worker Hall of Fame at http://hall.spacewalkoffame.org.

55

As the chart below shows, there are now 17 U. S. spaceports either already licensed by the Federal Aviation Administration (FAA) or with applications in process. **Commercial space business is being sought by all of them**, including space tourism and space-related manufacturing. **No matter where you live, you can join in.** See the websites listed in this booklet and at http://tinyurl.com/hwadknx.

U.S. Spaceports
Commercial/Government/Private Active and Proposed Launch Sites

- Poker Flat Research Range
- Kodiak Launch Complex
- California Spaceport
- Mojave Air and Space Port
- Edwards AFB
- Vandenberg AFB
- Spaceport America
- White Sands Missile Range
- Blue Origin Launch Site
- Oklahoma Spaceport
- Midland Spaceport
- SpaceX Launch Site (under construction)
- Mid-Atlantic Regional Spaceport
- Wallops Flight Facility
- Cecil Field Spaceport
- Kennedy Space Center
- Cape Canaveral Air Force Station
- Spaceport Florida
- Sea Launch Platform, Equatorial Pacific Ocean
- Reagan Test Site, Kwajalein Atoll, Marshall Islands

Key
- FAA-Licensed Non-Federal Launch Site
- U.S. Federal Launch Site
- Owned by University of Alaska Geophysical Institute
- Sole Site Operator

Other spaceports have been proposed for: Alabama, Colorado, Hawaii, and Texas.

Updated September 2014

Federal Aviation Administration

Many of these sites have active display areas providing access to their artifacts and their expert staff. If you live near one of these sites, look them up on the Internet and contact them to schedule a visit to see our space program first-hand. There is nothing like seeing space hardware that has actually flown into space and talking to engineers, scientists, and astronauts who have worked on launch teams, manned consoles in control rooms, and flown into space. Don't miss your chance to start your space journey today.

Make no mistake - Space is a global enterprise for science, technology, power, and prestige. The chart below shows **54 types of launch systems**, the countries involved and the kinds of competition that exist. Without an active and growing space program, we cannot maintain our leadership in space. You can play a part in making our space program part of your life. **Please make your voice heard!**

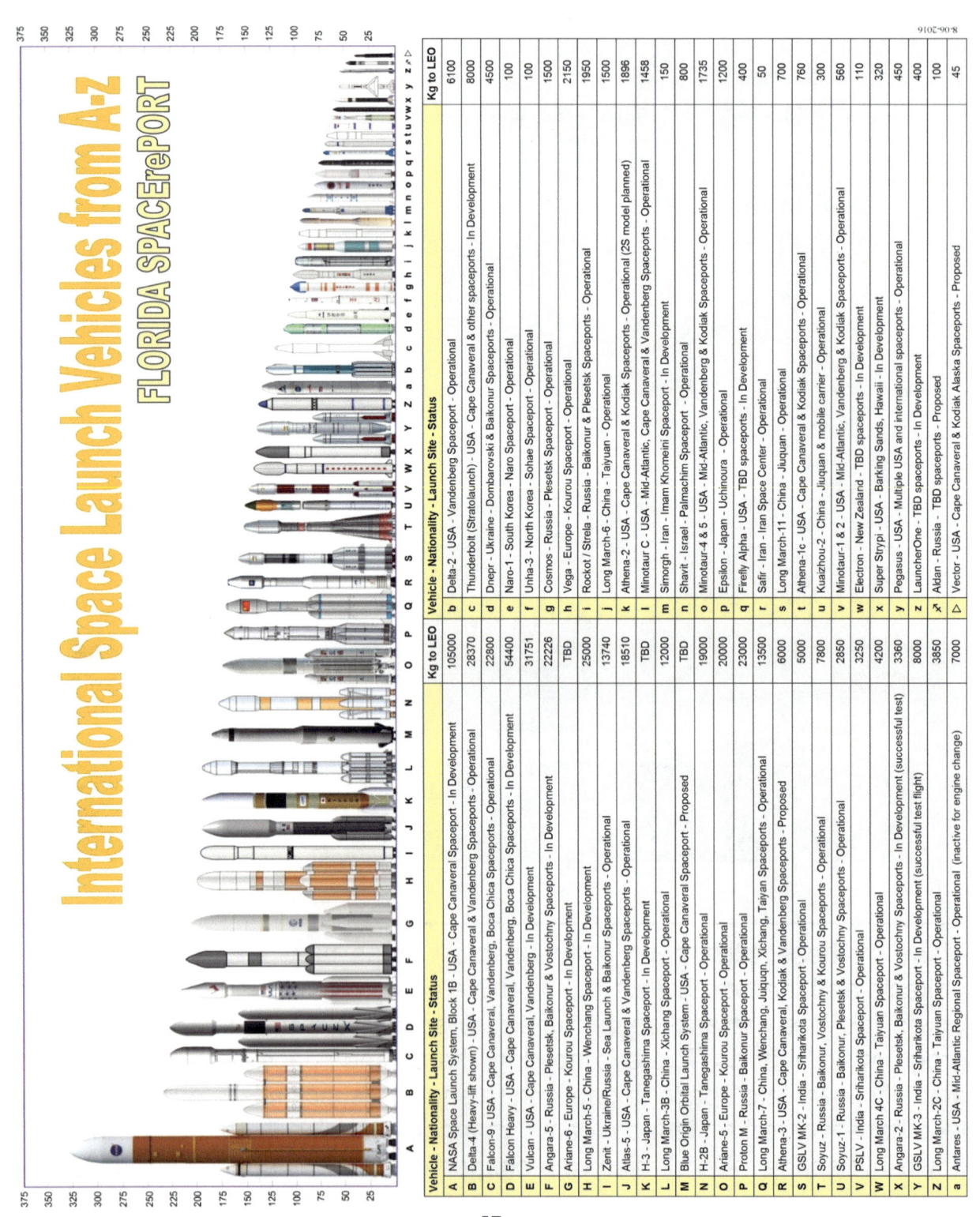

International Space Launch Vehicles from A-Z
FLORIDA SPACErePORT

	Vehicle - Nationality - Launch Site - Status	Kg to LEO
A	NASA Space Launch System, Block 1B - USA - Cape Canaveral Spaceport - In Development	105000
B	Delta-4 (Heavy-lift shown) - USA - Cape Canaveral & Vandenberg Spaceports - Operational	28370
C	Falcon-9 - USA - Cape Canaveral, Vandenberg, Boca Chica Spaceports - Operational	22800
D	Falcon Heavy - USA - Cape Canaveral, Vandenberg, Boca Chica Spaceports - In Development	54400
E	Vulcan - USA - Cape Canaveral, Vandenberg - In Development	31751
F	Angara-5 - Russia - Plesetsk, Baikonur & Vostochny Spaceports - In Development	22226
G	Ariane-6 - Europe - Kourou Spaceport - In Development	TBD
H	Ariane-5 - Europe - Kourou Spaceport - Operational	25000
I	Zenit - Ukraine/Russia - Sea Launch & Baikonur Spaceports - Operational	13740
J	Atlas-5 - USA - Cape Canaveral & Vandenberg Spaceports - Operational	18510
K	H-3 - Japan - Tanegashima Spaceport - In Development	TBD
L	Long March-3B - China - Xichang Spaceport - Operational	12000
M	Blue Origin Orbital Launch System - USA - Cape Canaveral Spaceport - Proposed	TBD
N	H-2B - Japan - Tanegashima Spaceport - Operational	19000
O	Ariane-5 - Europe - Kourou Spaceport - Operational	20000
P	Proton M - Russia - Baikonur Spaceport - Operational	23000
Q	Long March-7 - China, Wenchang, Juiquan, Xichang, Taiyan Spaceports - Operational	13500
R	Athena-3 - USA - Cape Canaveral, Kodiak & Vandenberg Spaceports - Proposed	6000
S	GSLV MK-2 - India - Sriharikota Spaceport - Operational	5000
T	Soyuz - Russia - Baikonur, Vostochny & Kourou Spaceports - Operational	7800
U	Soyuz-1 - Russia - Baikonur, Plesetsk & Vostochny Spaceports - Operational	2850
V	PSLV - India - Sriharikota Spaceport - Operational	3250
W	Long March 4C - China - Plesetsk, Baikonur & Vostochny Spaceports - Operational	4200
X	Angara-2 - Russia - Plesetsk, Baikonur & Vostochny Spaceports - In Development (successful test)	3360
Y	GSLV MK-3 - India - Sriharikota Spaceport - In Development (successful test flight)	8000
Z	Long March-2C - China - Taiyuan Spaceport - Operational	3850
a	Antares - USA - Mid-Atlantic Regional Spaceport - Operational (inactive for engine change)	7000

	Vehicle - Nationality - Launch Site - Status	Kg to LEO
b	Delta-2 - USA - Vandenberg Spaceport - Operational	6100
c	Thunderbolt (Stratolaunch) - USA - Cape Canaveral & other spaceports - In Development	8000
d	Dnepr - Ukraine - Dombarovski & Baikonur Spaceports - Operational	4500
e	Naro-1 - South Korea - Naro Spaceport - Operational	100
f	Unha-3 - North Korea - Sohae Spaceport - Operational	100
g	Cosmos - Russia - Plesetsk Spaceport - Operational	1500
h	Vega - Europe - Kourou Spaceport - Operational	2150
j	Rockot / Strela - Russia - Baikonur & Plesetsk Spaceports - Operational	1950
k	Long March-6 - China - Taiyuan - Operational	1500
k	Athena-2 - USA - Cape Canaveral & Kodiak Spaceports - Operational (2S model planned)	1896
l	Minotaur C - USA - Mid-Atlantic, Cape Canaveral & Vandenberg Spaceports - Operational	1458
m	Simorgh - Iran - Imam Khomeini Spaceport - In Development	150
n	Shavit - Israel - Palmachim Spaceport - Operational	800
o	Minotaur-4 & 5 - USA - Mid-Atlantic, Vandenberg & Kodiak Spaceports - Operational	1735
p	Epsilon - Japan - Uchinoura - Operational	1200
q	Firefly Alpha - USA - TBD spaceports - In Development	400
r	Safir - Iran - Iran Space Center - Operational	50
s	Long March-11 - China - Jiuquan - Operational	700
t	Athena-1c - USA - Cape Canaveral & Kodiak Spaceports - Operational	760
u	Kuaizhou-2 - China - Jiuquan & mobile carrier - Operational	300
v	Minotaur-1 & 2 - USA - Mid-Atlantic, Vandenberg & Kodiak Spaceports - Operational	560
w	Electron - New Zealand - TBD spaceports - In Development	110
x	Super Strypi - USA - Barking Sands, Hawaii - In Development	320
y	Pegasus - USA - Multiple USA and international spaceports - Operational	450
z	LauncherOne - USA - TBD spaceports - In Development	400
z²	Aldan - Russia - TBD spaceports - Proposed	100
Δ	Vector - USA - Cape Canaveral & Kodiak Alaska Spaceports - Proposed	45

WEB LINKS FOR ADDITIONAL INFORMATION

History and Space Technology (Aerospace 101)
1. http://www.solarviews.com/eng/rocket.htm (Brief History)
2. http://www.spaceline.org/rockethistory.html (History)
3. http://tinyurl.com/zgyewao (NASA Space Shuttle)
4. http://science.howstuffworks.com/rocket1.html (Motor theory)
5. http://octopus.gma.org/surfing/sats.html (Orbital dynamics)

General Information – Organizations, Sites, Missions, Hardware
1. http://www.flyrockets.com/clubs.asp (Fly Rockets!)
2. http://tinyurl.com/hwadknx (FAA Compendium 2016)
3. http://tinyurl.com/jfa7euk (Aerial Views of KSC and CCAFS)
4. http://tinyurl.com/o6z8cbq (International Space Station - ISS)
5. http://www.spacex.com/ (SpaceX)
6. http://www.spacex.com/falcon9 (Falcon 9)
7. http://www.spacex.com/dragon (Dragon Spacecraft)
8. http://www.spacex.com/falcon-heavy (Falcon Heavy)
9. https://www.blueorigin.com/technology (Blue Origin)
10. http://tinyurl.com/juvuttr (New Shepard Launch and Landing)
11. http://www.ulalaunch.com/ (United Launch Alliance)
12. http://www.ulalaunch.com/products_atlasv.aspx (Atlas V)
13. http://www.ulalaunch.com/products_deltaiv.aspx (Delta IV)
14. http://www.lockheedmartin.com/us/ssc/orion-eft1.html (Orion)
15. http://www.boeing.com/space/ (Boeing – Space)
16. http://tinyurl.com/q4e9s7m (CST-100 Starliner)
17. http://tinyurl.com/j8x3c3m (X-37B Orbital Test Vehicle)
18. http://tinyurl.com/jglbzkc (Sierra Nevada Corporation)
19. http://tinyurl.com/jz8wqjp (Dream Chaser Space Plane)
20. http://tinyurl.com/h4oe5qp (Moon Express At KSC)
21. http://oneweb.world/ (OneWeb Factory at KSC)
22. http://tinyurl.com/zwxp3ne (NASA Kennedy Space Center)
23. https://www.kennedyspacecenter.com/ (KSC Visitor Complex)
24. http://afspacemuseum.org/ccafs/ (AF Space Museum, CCAFS)
25. http://spacewalkoffame.org/ (U. S. Space Walk of Fame Museum)
26. http://hall.spacewalkoffame.org/ (U. S. Space Worker Hall of Honor)
27. http://tinyurl.com/zip2vms (What Kind of World Do You Want?)

ABOUT THE FLORIDA SPACEPORT

The Florida Spaceport comprises **two federal facilities** – the **16,238 acre Cape Canaveral Air Force Station** on the east adjacent to the ocean, and the **140,000 acre John F. Kennedy Space Center** on Merritt Island, in the center of the map. Together they host more than $15B in space launch facilities: hangars, test bays, assembly areas, labs and shops, launch pads, rail systems, waterways and piers, medical offices and life sciences facilities, administrative offices, specialized servicing areas including two landing fields and storage areas for fuels and ordnance items used for launch. KSC is home to the **Merritt Island National Wildlife Refuge**, and the Cape is home to **Canaveral Light** – one of the oldest lighthouses in the U.S.

ABOUT THE AUTHOR

Albert Koller, Jr. was born in Baltimore, Maryland, October 7, 1941 and moved to Florida in 1958. He began his aerospace career at age 17 working summers for the **Army Ballistic Missile Agency** (ABMA) at its Missile Firing Lab on Cape Canaveral as part of the **Von Braun ABMA rocket team**. In 1960 he was transferred to NASA and was a member of the **Launch Vehicle team** for the Apollo program, manning a Firing Room console for **Apollo 11**. He remained at NASA for 32 years and took early retirement in 1992 to pursue a lifelong commitment to education. In 2001 he led the creation of ***SpaceTEC®***, the **National Science Foundation's Center for Aerospace Technical Education**.

Dr. Koller is a certified professional manager (CPM) and holds degrees in math and physics (BA), systems management (MS) and business administration (DBA). He consults and is nationally published. He was the 2001 recipient of the Florida Space Business Roundtable's **"Explorer Award"** for his leadership in space-related education. The National Space Club Florida Committee awarded him its **Lifetime Achievement Award in Aerospace** in August 2008. In October, 2011 he received the **Thomas E. Gamble Excellence in Education for Economic Development Award**. He retired in March, 2013 and was awarded a ***SpaceTEC®* certificate** as an **Honorary Certified Aerospace Technician**. He is married to the former Carol Ann Knight of Jacksonville, Florida, and they have 3 daughters, a son, 10 grandchildren, and 4 great grandchildren.

Made in the USA
Columbia, SC
26 June 2025